MATHEMATICAL
MINDSETS

JO BOALER

FOREWORD BY CAROL DWECK

MATHEMATICAL MINDSETS

Unleashing Students' POTENTIAL Through
Creative Math, Inspiring Messages and
INNOVATIVE TEACHING

JOSSEY-BASS™

A Wiley Brand

Published by Jossey-Bass
A Wiley Brand
One Montgomery Street, Suite 1000, San Francisco, CA 94104-4594—www.josseybass.com

Jossey-Bass books and products are available through most bookstores. To contact Jossey-Bass directly call our Customer Care Department within the U.S. at 800-956-7739, outside the U.S. at 317-572-3986, or fax 317-572-4002.

Wiley publishes in a variety of print and electronic formats and by print-on-demand. Some material included with standard print versions of this book may not be included in e-books or in print-on-demand. If this book refers to media such as a CD or DVD that is not included in the version you purchased, you may download this material at http://booksupport.wiley.com. For more information about Wiley products, visit www.wiley.com.

Library of Congress Cataloging-in-Publication Data
Boaler, Jo, 1964–
  Mathematical mindsets : unleashing students' potential through creative math, inspiring messages, and innovative teaching / Jo Boaler.
    1 online resource.
    "A Wiley brand."
    Includes bibliographical references and index.
    Description based on print version record and CIP data provided by publisher; resource not viewed.
    ISBN 978-1-118-41827-7 (pdf) – ISBN 978-1-118-41553-5 (epub) – ISBN 978-0-470-89452-1 (pbk.)
  1. Mathematics–Study and teaching (Middle school) 2. Mathematics–Study and teaching (Secondary) I. Title.
  QA135.6
  510.71'2–dc23

                                                                                            2015031316

Cover Design: Wiley
Cover Image: Illustration © agsandrew/Shutterstock; Schoolchildren © monkeybusinessimages/iStockphoto.com

Printed in the United States of America

FIRST EDITION

V10010753_060319

Contents

For Jaime and Ariane, my two girls who inspire me every day.

FOREWORD

One of my former Stanford students teaches fourth grade in the South Bronx, an area of New York City with many underserved, underachieving minority students. Her students invariably believe they are bad at math, and if you looked at their past performance, you might be tempted to think so too. And yet, after one year in her class, her fourth graders became the #1 fourth-grade class in the state of New York: 100% of them passed the state math test, with 90% of them earning the top score. And this is just one of many examples of how all students can learn math.

When people think that some kids just can't do math, that success in math is reserved for only certain kids, thought of as "smart," or that it's just too late for kids who haven't had the right background, then they can easily accept that many students fail math and hate math. In fact, we have found that many teachers actually console their students by telling them not to worry about doing poorly in math because not everyone can excel in it. These adult enablers—parents and teachers alike—allow kids to give up on math before they've barely gotten started. No wonder more than a few students simply dismiss their own poor performance by declaring: "I'm not a math person."

Where do parents, teachers, and students get the idea that math is just for some people? New research shows that this idea is deeply embedded in the field of mathematics. Researchers polled scholars (at American universities) in a range of disciplines. They asked them how much they thought that success in their field depended on fixed, innate ability that cannot be taught, as opposed to hard work, dedication, and learning. Of all the STEM fields (science, technology, engineering, and math), math scholars were the most extreme in emphasizing fixed, innate ability (Leslie, Cimpian, Meyer, & Freeland, 2015). Other researchers are finding that many math instructors begin their courses by referring to students who have the aptitude and those who do not. One college instructor, on the first day of an introductory college course, was heard to say, "If it's not easy for you, you don't belong here" (Murphy, Garcia, & Zirkel, in prep). If this message is passed down from generation to generation, no wonder students are afraid of math. And no wonder they conclude they're not math people when it doesn't come easily.

But when we begin to see evidence that most students (and maybe almost all students) are capable of excelling in and enjoying math, as the following chapters show, it is no longer acceptable that so many students fail math and hate math. So what can we do to make math learning happen for all students? How can we help teachers and children believe that math ability can be developed, and then show teachers how to teach math in a way that brings this belief to life? That's what this book is about.

In this unique and wonderful book, Jo Boaler distills her years of experience and her powerful wisdom to show teachers exactly how to present math work, structure math problems, guide students through them, and give feedback in a way that helps students toward a "growth mindset" and keeps them there. Boaler is one of those rare and remarkable educators who not only know the secret of great teaching but also know how to give that gift to others. Thousands of teachers have learned from her, and here's what they say:

> "Throughout my schooling years … I was left feeling stupid and incapable of doing [math] … I cannot tell you the relief I now have that I can learn math myself, and I can teach students that they can too."

"[You have] helped me think about the transition to common core and how to help my students develop a love and curiosity for math."

"I was searching for a process of learning math that would change the attitude of students from dislike to enjoy … this was the change I needed."

Imagine your students joyfully immersed in really hard math problems. Imagine them begging to have their mistakes discussed in front of the class. Imagine them saying, "I *am* a math person!" This utopian vision is happening in classrooms around the world, and as you follow the advice in this book, you may well see it happening in your classroom too.

Carol Dweck
Professor of psychology and author of
Mindset: The New Psychology of Success

INTRODUCTION: THE POWER OF MINDSET

I remember clearly the fall afternoon that I sat down with my dean in her office, waiting for what would turn out to be a very important meeting. I had only recently returned to Stanford University from England where I was the Marie Curie Professor for Mathematics Education. I was still getting used to the change from the grey cloudy skies that seemed to be my constant companion during the three years I was on the Sussex coast in England to the sunshine that shines down on Stanford's campus almost continuously. I walked into the dean's office that day with some anticipation, as I was going to meet Carol Dweck for the first time. I was a little nervous to meet the famous researcher whose books on mindset had revolutionized people's lives, across continents, and whose work had moved governments, schools, parents, and even leading sports teams to approach life and learning differently.

Carol and her research teams have collected data over many years that support a clear finding—that everyone has a mindset, a core belief about how they learn (Dweck, 2006b). People with a growth mindset are those who believe that smartness increases with hard work, whereas those with a fixed mindset believe that you can learn things but you can't change your basic level of intelligence. Mindsets are critically important because research has shown that they lead to different learning behaviors, which in turn create different learning outcomes for students. When people change their mindsets and start to believe that they can learn to high levels, they change their learning pathways (Blackwell, Trzesniewski, & Dweck, 2007) and achieve at higher levels, as I will share in this book.

In our conversation that day, I asked Carol if she had thought about working with mathematics teachers, as well as students, because I knew that mindset interventions given to students help them, but math teachers have the potential to deeply impact students' learning in a sustained way over time. Carol responded enthusiastically and agreed with me that math was the subject most in need of a mindset makeover. That was the first of what would become many enjoyable conversations and collaborations over the next four years, which now include our working together on shared research projects with math teachers and presenting our research and ideas to them in workshops. My work on mindset and math over recent years has helped me develop a deep appreciation of the need to teach students about mindset *inside* mathematics, rather than in general. Students have such strong and often negative ideas about math that they can develop a growth mindset about everything else in their life but still believe that you can either achieve highly in math or you can't. To change these damaging beliefs, students need to develop *mathematical mindsets*, and this book will teach you ways to encourage them.

The fixed mindsets that many people hold about mathematics often combine with other negative beliefs about mathematics, to devastating effect. This is why it is so important to share with learners the new knowledge we have of mathematics and learning that I set out in this book. Recently I shared some of the ideas in an online class for teachers and parents—what has come to be known as a MOOC (massive, open, online class)—and the results were staggering, surpassing even my highest expectations (Stanford Center for Professional Development, n.d.).

Over forty thousand people enrolled in the class—teachers of all grade levels, and parents—and at the end, 95% of them said they would change their teaching or ways of helping their own children, because of the new knowledge they had learned. Additionally, over 65% of participants stayed in the course, not the 5% that MOOCS typically retain. The amazing response to my course came about because our new knowledge of the brain and mathematics learning is incredibly powerful and important.

When I taught my online class, and I read all the responses from the people who took it, I realized more strongly than ever before that many people have been traumatized by math. Not only did I find out how widespread the trauma is, but the evidence I collected showed that the trauma is fuelled by incorrect beliefs about mathematics and intelligence. Math trauma and math anxiety is kept alive within people because these incorrect beliefs are so widespread that they permeate society in the United States, the United Kingdom, and many other countries in the world.

I first became aware of the extent of math trauma in the days after I released my first book for parents and teachers, titled *What's Math Got to Do with It* in the United States and *The Elephant in the Classroom* in the United Kingdom. That book details the teaching and parenting changes we need to make for math to be more enjoyable and achievable. After the book was released, I was invited onto numerous different radio shows, on both sides of the Atlantic, to chat with the hosts about mathematics learning. These varied from breakfast show chats to a 20-minute, in-depth discussion with a very thoughtful PBS host and a spot on a much-loved British radio show called *Women's Hour*. Talking with radio hosts was a really interesting experience. I started most of the conversations talking about the changes we need to make, pointing out that math is traumatic for many people. This statement seemed to relax the hosts and caused many of them to open up and share with me their own stories of math trauma. Many of the interviews then turned into what seemed like therapy sessions, as the highly accomplished and knowledgeable professionals shared their various tales of math trauma, usually triggered by something a single math teacher had said or done. I still remember Kitty Dunne in Wisconsin telling me that the name of her algebra book was "burned" into her brain, revealing the strength of the negative associations she held onto. Jane Garvey at the BBC, an amazing woman for whom I have complete admiration, told me that she was so scared of mathematics that she had been fearful of interviewing me, and she had already told her two daughters that she was terrible at mathematics in school (something you should never do, as I will discuss later). This level of intensity of negative emotion around mathematics is not uncommon. Mathematics, more than any other subject, has the power to crush students' spirits, and many adults do not move on from mathematics experiences in school if they are negative. When students get the idea they cannot do math, they often maintain a negative relationship with mathematics throughout the rest of their lives.

Mathematics trauma does not reside only in people in the arts or entertainment professions. The release of my books led to meetings with some incredible people, one of the most interesting of whom was Dr. Vivien Perry. Vivien is a top scientist in England; she was recently awarded an OBE, the greatest honor bestowed in England, given by the queen. Her list of accomplishments is long, including being the vice chair of council for University College, London; a member of the medical research council; and a presenter of BBC TV science programs. Surprisingly perhaps, with Vivien's scientific career, she talks publicly and openly about a crippling fear of mathematics. Vivien has shared with me that she is so scared of mathematics that she cannot work out percentages when she needs to complete tax documents at home. In the months before I left the United Kingdom and returned to Stanford University, I presented at the Royal Institution in London. This was a great honor, to present at one of Britain's oldest and most respected institutions that has the worthy goal of bringing scientific work to the public. Every year in Britain the Christmas Lectures, founded by Michael Faraday in 1825, are aired on TV, given by eminent scientists who share their work with the public. I had asked Vivien to introduce me at the Royal Institution, and during that introduction she shared with the audience that when she was a child she had been made to stand in the corner by her mathematics teacher, Mrs. Glass, for not being able to recite her seven times table. She then went on to make the audience laugh by telling them that when she shared this story on the BBC, six women called the BBC action line and asked—was it Mrs. Glass of Boxbury School? Vivien shared that indeed it was.

Fortunately, such harsh teaching practices are almost extinct, and I continue to be inspired by the devotion and commitment of most mathematics teachers I work with. But we know that negative and damaging messages are still handed out to students every day—messages that are not intended to harm, but that we know can start students on a damaging and lasting mathematics pathway. Such pathways can be reversed, at any time, but for many they are not, and they affect every future experience of mathematics that people have. Changing the messages that students receive about mathematics is not, sadly, as simple as just changing the words teachers and parents use, although words are very important. Students also receive and absorb many indirect messages about mathematics through many aspects of math teaching, such as the questions they work on in math class, the feedback they get, the ways they are grouped, and other aspects of mathematics teaching and help that we will consider together in this book.

Vivien is convinced that she has a brain condition, called dyscalculia, that stops her from being successful with math. But we now know that one experience or message can change everything for students (Cohen & Garcia, 2014), and it seems very likely that Vivien's negative math experiences were at the root of the math anxiety she now struggles with daily. Vivien—fortunately for the many who have benefited from her work—was able to be successful despite her mathematics experiences, even in a quantitative field, but most people are not so fortunate, and the early damaging experiences they have with mathematics close doors for them for the rest of their lives.

Taking math courses matters. Research studies have established that the more math classes students take, the higher their earnings ten years later, with advanced math courses predicting an increase in salary as high as 19.5% ten years after high school (Rose & Betts, 2004). Research has also found that students who take advanced math classes learn ways of working and thinking—especially learning to reason and be logical—that make them more productive in their jobs. Students taking advanced math learn how to approach mathematical situations so that once they are employed, they are

(continued)

(*continued*)

promoted to more demanding and more highly paid positions than those who did not take mathematics to advanced levels (Rose & Betts, 2004). In my study of schools in England, I found that students were advanced in their jobs, ending up with higher-paid employment, because they learned mathematics through a project-based approach in high school that I will discuss in later chapters (Boaler, 2005).

We all know that math trauma exists and is debilitating for people; numerous books have been devoted to the subject of math anxiety and ways to help people overcome it (Tobias, 1978). It would be hard to overstate the number of people who walk on our planet who have been harmed by bad math teaching, but the negative ideas that prevail about math do not come only from harmful teaching practices. They come from one idea, which is very strong, permeates many societies, and is at the root of math failure and underachievement: that only some people can be good at math. That single belief—that math is a "gift" that some people have and others don't—is responsible for much of the widespread math failure in the world.

So where does that damaging idea—an idea that notably is absent in countries such as China and Japan that top the world in math achievement—come from? I am fortunate enough to have two daughters who at the time of this writing are in third and sixth grades in California. This means that I now have the dubious pleasure of catching regular glimpses of "tweenie" TV programs. This has been very enlightening—and worrying—as a day does not go by when mathematics doesn't comes up in one of these TV programs in a negative light. Math is conveyed as a really hard subject that is uninteresting, inaccessible, and only for "nerds"; it is not for cool, engaging people, and it is not for girls. It is no wonder that so many children in schools disengage from math and believe they cannot do well.

The idea that only some people can do math is embedded deep in the American and British psyche. Math is special in this way, and people have ideas about math that they don't have about any other subject. Many people will say that math is different because it is a subject of right and wrong answers, but this is incorrect, and part of the change we need to see in mathematics is acknowledgment of the creative and interpretive nature of mathematics. Mathematics is a very broad and multidimensional subject that requires reasoning, creativity, connection making, and interpretation of methods; it is a set of ideas that helps illuminate the world; and it is constantly changing. Math problems should encourage and acknowledge the different ways in which people see mathematics and the different pathways they take to solve problems. When these changes happen, students engage with math more deeply and well.

Another misconception about mathematics that is pervasive and damaging—and wrong—is the idea that people who can do math are the smartest or cleverest people. This makes math failure particularly crushing for students, as they interpret it as meaning that they are not smart. We need to dispel this myth. The combined weight of all the different wrong ideas about math that prevail in society is devastating for many children—they believe that mathematics ability is a sign of intelligence and that math is a gift, and if they don't have that gift then they are not only bad at math but they are unintelligent and unlikely to ever do well in life.

As I write this book, it is clear that the world is developing a great appreciation for and understanding of the importance of mindset. Carol Dweck's book has been translated into more than 20 languages (Dweck, 2006b), and interest in the impact of mindset continues to grow. What is less well known is how mindset ideas are infused through all of mathematics, and how teachers of mathematics and parents working with their students at home can transform students' ideas, experiences, and life chances through a growth mindset approach to math. General mindset interventions can be helpful for shifting students' mindsets, but if students return to mathematics classrooms and math work at home working in the same ways they always have, that growth mindset about math slowly erodes away. The ideas that I share with teachers and parents and set out in this book include paying attention to the math questions and tasks that students work on, the ways teachers and parents encourage or grade students, the forms of grouping used in classrooms, the ways mistakes are dealt with, the norms developed in classrooms, the math messages we can give to students, and the strategies they learn to approach math—really, the whole of the mathematics teaching and learning experience. I am excited to share this new knowledge with you, and I am confident it will help you and anyone you work with on mathematics.

In the next chapter I will set out some of the fascinating and important ideas that have emerged from research in recent years; in the eight chapters that follow, I will focus on the strategies that can be used in math classrooms and homes to implement the ideas I share in these first two chapters. I strongly recommend reading all of the chapters, skipping to the strategies will not be helpful if the underlying ideas are not well understood.

In the months after my online MOOC was released to teachers and parents, I received thousands of letters, emails, and other messages from people sharing with me the changes they had made in their classrooms and homes and the impact this had on the students. Relatively small changes in teaching and parenting can change students' mathematical pathways, because the new knowledge we have on the brain, mindset, and mathematics learning is truly revolutionary. This book is about the creation of *mathematical mindsets* through a new kind of teaching and parenting that is, at its heart, about growth, innovation, creativity, and the fulfillment of mathematics potential. Thank you for joining me, and for taking steps on a pathway that could change your and your students' relationships with mathematics forever.

MATHEMATICAL
MINDSETS

CHAPTER 1

The Brain and Mathematics Learning

In the last decade we have seen the emergence of technologies that have given researchers new access into the workings of the mind and brain. Now scientists can study children and adults working on math and watch their brain activity; they can look at brain growth and brain degeneration, and they can see the impact of different emotional conditions upon brain activity. One area that has emerged in recent years and stunned scientists concerns "brain plasticity." It used to be believed that the brains people were born with couldn't really be changed, but this idea has now been resoundingly disproved. Study after study has shown the incredible capacity of brains to grow and change within a really short period (Abiola & Dhindsa, 2011; Maguire, Woollett, & Spiers, 2006; Woollett & Maguire, 2011).

When we learn a new idea, an electric current fires in our brains, crossing synapses and connecting different areas of the brain (see Figure 1.1).

If you learn something deeply, the synaptic activity will create lasting connections in your brain, forming structural pathways, but if you visit an idea only once or in a superficial way, the synaptic connections can "wash away" like pathways made in the sand. Synapses fire when learning happens, but learning does not happen only in classrooms or when reading books; synapses fire when we have conversations, play games, or build with toys, and in the course of many, many other experiences.

A set of findings that caused scientists to change what they thought about ability and learning came from research on the brain growth shown by Black Cab drivers in London. I am from England, and I have travelled in taxicabs in London many times. I still have fond memories of the exciting day trips my family and I took to London when I was a child, from our home a few hours away. As an adult I studied and worked at King's College, London University, and had many more opportunities for trips around London in taxis. A number of different taxis work in the London area, but the queen bee of taxis in London is the "Black Taxi" or "Black Cab" (see Figure 1.2).

For most of my rides through London in a Black Cab I had no idea how highly qualified the drivers were. It turns out that to become a Black Cab driver in London,

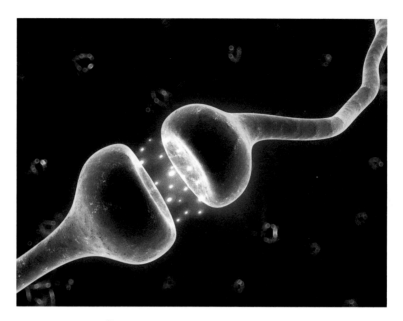

FIGURE 1.1   A synapse fires

FIGURE 1.2   The Black Cab of London

FIGURE 1.3   Map of London

applicants need to study for two to four years and during that time memorize an incredible 25,000 streets and 20,000 landmarks within a 25-mile radius of Charing Cross in London. Learning your way around the city of London is considerably more challenging than learning your way around most American cities, as London is not built on a grid structure and comprises thousands of interweaving, interconnected streets (see Figure 1.3).

At the end of their training period the Black Cab drivers take a test that is simply and elegantly called "The Knowledge." If you ride in a London Black Cab and ask your driver about "The Knowledge," they are usually happy to regale you with stories of the difficulty of the test and their training period. The Knowledge is known to be one of the world's most demanding courses, and applicants take the test an average of 12 times before passing.

In the early 2000s scientists chose to study London Black Cab drivers to look for brain changes as the drivers took years of complex spatial training, but the scientists were not

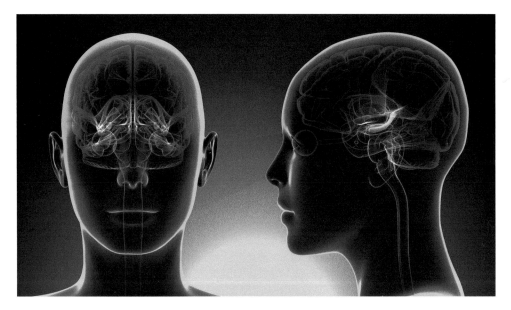

FIGURE 1.4   The hippocampus

expecting such dramatic results. Researchers found that at the end of the training period the hippocampus in the taxi drivers' brains had grown significantly (Maguire et al., 2006; Woollett & Maguire, 2011). The hippocampus is the brain area specialized in acquiring and using spatial information (see Figure 1.4).

In other studies, scientists compared the brain growth of Black Cab drivers to that of London bus drivers. Bus drivers learn only simple and singular routes, and the studies showed that they did not experience the same brain growth (Maguire et al., 2006). This confirmed the scientists' conclusion that the Black Cab drivers' unusually complex training was the reason for their dramatic brain growth. In a further study, scientists found that after Black Cab drivers retired, their hippocampus shrank back down again (Woollett & Maguire, 2011).

The studies conducted with Black Cab drivers, of which there have now been many (Maguire et al., 2006; Woollett & Maguire, 2011), showed a degree of brain flexibility, or plasticity, that stunned scientists. They had not previously thought that the extent of brain growth they measured was possible. This led to a shift in the scientific world in thinking about learning and "ability" and the possibility of the brain to change and grow.

Around the time that the Black Cab studies were emerging, something happened that would further rock the scientific world. A nine-year old girl, Cameron Mott, had been having seizures that the doctors could not control. Her physician, Dr. George Jello, proposed something radical. He decided he should remove half of her brain, the entire left hemisphere. The operation was revolutionary—and ultimately successful. In the days following her operation, Cameron was paralyzed. Doctors expected her to be disabled for many years, as the left side of the brain controls physical movements. But as weeks and months passed, she stunned doctors by recovering function and movement that could mean only one thing—the right side of her brain was developing the connections it needed to perform the functions of the left side of the brain. Doctors attributed this to the incredible plasticity of the brain and could only conclude that the brain had, in effect, "regrown." The new brain growth had happened faster than doctors imagined possible. Now Cameron runs and plays with other children, and a slight limp is the only sign of her significant brain loss (http://www.today.com/id/36032653/ns/today-today_health/t/meet-girl-half-brain/# .UeGbixbfvCE)

The new findings that brains can grow, adapt, and change shocked the scientific world and spawned new studies of the brain and learning, making use of ever-developing new technologies and brain scanning equipment. In one study that I believe is highly significant for those of us in education, researchers at the National Institute for Mental Health gave people a 10-minute exercise to work on each day for three weeks. The researchers compared the brains of those receiving the training with those who did not. The results showed that the people who worked on an exercise for a few minutes each day experienced structural brain changes. The participants' brains "rewired" and grew in response to a 10-minute mental task performed daily over 15 weekdays (Karni et al., 1998). Such results should prompt educators to abandon the traditional fixed ideas of the brain and learning that currently fill schools—ideas that children are smart or dumb, quick or slow. If brains can change in three weeks, imagine what can happen in a year of math class if students are given the right math materials and they receive positive messages about their potential and ability. Chapter Five will explain the nature of the very best mathematics tasks that students should be working on to experience this brain growth.

The new evidence from brain research tells us that everyone, with the right teaching and messages, can be successful in math, and everyone can achieve at the highest levels in school. There are a few children who have very particular special educational needs that make math learning difficult, but for the vast majority of children—about 95%—any levels of school math are within their reach. And the potential of the brain to grow and change is just as strong in children with special needs. Parents and teachers need to know this important information. When I share this evidence with teachers in workshops and presentations, most of them are encouraged and inspired, but not all of them. I was with a group of teachers recently, and one high school math teacher was clearly troubled by the idea. He said, "You aren't telling me, are you, that *any* of the sixth graders in my school could take calculus in twelfth grade?" When I said, "That is exactly what I am saying," I could tell he was genuinely troubled by the idea—although, to his credit, he was not rejecting it outright. Some teachers find the idea that anyone can learn math to high levels difficult to accept, especially if they have spent many years deciding who can and who can't do math and teaching them accordingly. Of course, sixth graders have had many experiences and messages since birth that have held some of them back, and some students may come to sixth grade with significantly less mathematical knowledge than others, but this doesn't mean they cannot accelerate and reach the highest levels—they can, if they receive the high-quality teaching and support that all children deserve.

I am often asked whether I am saying that everyone is born with the same brain. I am not. What I am saying is that any brain differences children are born with are nowhere near as important as the brain growth experiences they have throughout life. People hold very strong views that the way we are born determines our potential; they point to well-known people who were considered geniuses—such as Albert Einstein or Ludwig van Beethoven. But scientists now know that any brain differences present at birth are eclipsed by the learning experiences we have from birth onward (Wexler in Thompson, 2014). Every second of the day our brain synapses are firing, and students raised in stimulating environments with growth mindset messages are capable of anything. Brain differences can give some people a head start, but infinitesimally small numbers of people have the sort of head start that gives them advantages over time. And those people who are heralded as natural geniuses are the same people who often stress the hard work they have put in and the number of mistakes they made. Einstein, probably the most well known of those thought to be a genius, did not learn to read until he was nine and spoke often about his achievements coming from the number of mistakes he had made and the persistence he had shown. He tried hard, and when he made mistakes he tried harder. He approached work and life with the attitude of someone with a growth mindset. A lot of scientific evidence suggests that the difference between those who succeed and those who don't is not the brains they were born with, but their approach to life, the messages they receive about their potential, and the opportunities they have to learn. The very best opportunities to learn come about when students believe in themselves. For far too many students in school, their learning is hampered by the messages they have received about their own potential, making them believe they are not as good as others, that they don't have the potential of others. This book provides the information you need, whether you are a teacher or parent, to give students the self-belief they need and should have; to set them on a pathway that leads to a mathematical mindset, whatever their prior experiences. This new pathway involves a change in the way students consider themselves and also a change in the way they approach the subject of mathematics, as the rest of the book will describe.

Although I am not saying that everyone is born with the same brain, I *am* saying that there is no such thing as a "math brain" or a "math gift," as many believe. No one is born knowing math, and no one is born lacking the ability to learn math. Unfortunately, ideas of giftedness are widespread. Researchers recently investigated the extent to which college professors held ideas about giftedness in their subject, and they found something remarkable (Leslie, Cimpian, Meyer, & Freeland, 2015). Math was the subject whose professors were found to hold the most fixed ideas about who could learn. Additionally, researchers found that the more a field values giftedness, the fewer female PhDs there were in the field and that field specific beliefs were correlated with female representation across all 30 fields they investigated. The reason that there are fewer women in fields where professors believe that only the "gifted" can achieve is that stereotypical beliefs still prevail about who really belongs, as chapter 6 describes. It is imperative for our society that we move to a more equitable and informed view of mathematics learning in our conversations and work with students. Our conversations and work need to reflect the new science of the brain and communicate to all that everyone can learn math well, not only those believed to hold a "gift". This could well be the key to unlocking a different future – one in which math trauma is a thing of the past and students from all backgrounds are given access to high quality mathematics learning opportunities.

In studies by Carol Dweck and her colleagues, about 40% of the children held a damaging fixed mindset, believing that intelligence is a gift that you either have or you don't. Another 40% of the students had a growth mindset. The remaining 20% wavered between the two mindsets

(Dweck, 2006b). Students with a fixed mindset are more likely to give up easily, whereas students with a growth mindset keep going even when work is hard and are persistent, displaying what Angela Duckworth has termed "grit" (Duckworth & Quinn, 2009). In one study, seventh-grade students were given a survey to measure their mindset, then researchers followed the students over two years to monitor their mathematics achievement. The results were dramatic, as the achievement of the students with a fixed mindset stayed constant, but the achievement of those with a growth mindset went onward and upward (Blackwell et al., 2007) (see Figure 1.5).

In other studies, researchers have shown that students' (and adults') mindsets can change from fixed to growth, and when that happens their learning approach becomes significantly more positive and successful (Blackwell et al., 2007). We also have new evidence, that I review in Chapter Two, that students with a growth mindset have more positive brain activity when they make mistakes, with more brain regions lighting up and more attention to and correcting of errors (Moser, Schroder, Heeter, Moran, & Lee, 2011).

I didn't need more evidence of the importance of helping students—and adults—develop a growth mindset in relation to math in particular, but recently I found myself sitting with the Program for International Student Assessment (PISA) team at the Organisation for Economic Co-operation and Development (OECD) in Paris, exploring with them their incredible data set of 13 million students worldwide. The PISA team gives international tests every four years, and the results are reported in news outlets across the globe. The test scores often start alarm bells ringing around the United States, and for good reason. In the latest tests, the United States ranked 36th out of 65 OECD countries in math performance (PISA, 2012)—a result that speaks, as do many others, to the incredible need to reform mathematics teaching and learning in the United States. But the PISA team not only administer math tests; they also survey students to collect their ideas and beliefs about mathematics and their mindsets. I was invited to work with

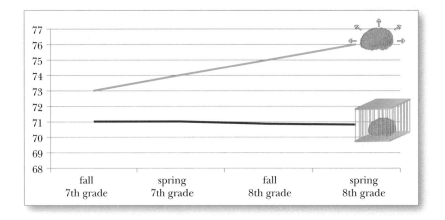

FIGURE 1.5 Students with a growth mindset outperform those with a fixed mindset in mathematics
Source: Blackwell et al., 2007.

the PISA team after some of the group took the online class I taught last summer. One of them was Pablo Zoido, a soft-spoken Spaniard who thinks deeply about math learning and has considerable expertise in working with giant data sets. Pablo is an analyst for PISA, and as he and I explored the data we saw something amazing—that the highest-achieving students in the world are those with a growth mindset, and they outrank the other students by the equivalent of more than a year of mathematics (see Figure 1.6).

The fixed mindset thinking that is so damaging—a mindset in which students believe they either are smart or are not—cuts across the achievement spectrum, and some of the students most damaged by these beliefs are high-achieving girls (Dweck, 2006a). It turns out that even believing you *are* smart—one of the fixed mindset messages—is damaging, as students with this fixed mindset are less willing to try more challenging work or subjects because they are afraid of slipping up and no longer being seen as smart. Students with a growth mindset take on hard work, and they view mistakes as a challenge and motivation to do more. The high incidence of fixed mindset thinking among girls is one reason that girls opt out of STEM subjects—science, technology, mathematics, and engineering. This not only reduces their own life chances but also impoverishes the STEM disciplines that need the thinking and perspectives that girls and women bring (Boaler, 2014a).

One reason so many students in the United States have fixed mindsets is the praise they are given by parents and teachers. When students are given fixed praise—for example, being told they are smart when they do something well—they may feel good at first, but when they fail later (and everyone does) they think that means they are not so smart after all. In a recent study, researchers found that the praise parents gave their babies between birth and age three predicted their mindsets five years later (Gunderson et al., 2013). The impact of the praise students receive can be so strong that it affects their behavior immediately. In one of Carol's studies, researchers asked

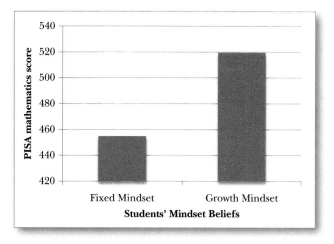

FIGURE 1.6 Mindset and mathematics
Source: PISA, 2012.

400 fifth graders to take an easy short test, on which almost all performed well. Half the children were then praised for "being really smart." The other half were complimented on "having worked really hard." The children were then asked to take a second test and choose between one that was pretty simple, that they would do well on, or one that was more challenging, that they might make mistakes on. Ninety percent of those who were praised for their effort chose the harder test. Of those praised for being smart, the majority chose the easy test (Mueller & Dweck, 1998).

Praise feels good, but when people are praised for who they are as a person ("You are so smart") rather than what they did ("That is an amazing piece of work"), they get the idea that they have a fixed amount of ability. Telling students they are smart sets them up for problems later. As students go through school and life, failing at many tasks—which, again, is perfectly natural—they evaluate themselves, deciding how smart or not smart this means they really are. Instead of praising students for being smart, or any other personal attribute, it's better to say things like: "It is great that you have learned that," and "You have thought really deeply about this."

Our education systems have been pervaded with the traditional notion that some students are not developmentally ready for certain levels of mathematics. A group of high school math teachers in a school I recently encountered had, shockingly, written to the school board arguing that some students could never pass algebra 2. They particularly cited minority students from low-income homes; they argued that these students could not learn algebra unless the teachers watered down the curriculum. Such deficit and racist thinking needs to be banished from schools. The letter written by the teachers was published in local newspapers and ended up being used in the state legislature as an example of the need for charter schools (Noguchi, 2012). The letter shocked many people, but unfortunately this idea that some students cannot learn high-level mathematics is shared by many. Deficit thinking can take all sorts of forms and is sometimes used with genuine concern for students—many people believe there is a developmental stage students must go through before they are ready for certain mathematics topics. But these ideas are also outdated, as students are as ready as the experiences they have had, and if students are not ready, they can easily become so with the right experiences, high expectations from others, and a growth mindset. There is no preordained pace at which students need to learn mathematics, meaning it is *not* true that if they have not attained a certain age or emotional maturity they cannot learn some mathematics. Students may be unready for some mathematics because they still need to learn some foundational, prerequisite mathematics they have not yet learned, but *not* because their brain cannot develop the connections because of their age or maturity. When students need new connections, they can learn them.

For many of us, appreciating the importance of mathematical mindsets and developing the perspective and strategies to change students' mindsets involves some careful thinking about our own learning and relationship with mathematics. Many of the elementary teachers I have worked with, some of whom took my online class, have told me that the ideas I gave them on the brain, on potential, and on growth mindsets has been life-changing for them. It caused them to develop a growth mindset in mathematics, to approach mathematics with confidence and enthusiasm and to pass that on to their students. This is often particularly important for elementary teachers, because many have, at some point in their own learning, been told *they* cannot do mathematics or that mathematics is not for them. Many teach mathematics with their own fear of the subject. The research I shared with them helped banish that fear and put them on a different mathematical journey. In an important study, Sian Beilock and colleagues found that the extent of negative emotions elementary teachers held about mathematics predicted the achievement of girls in their classes, but not boys (Beilock, Gunderson, Ramirez, & Levine, 2009). This gender difference probably comes about because girls identify with their female teachers, particularly in elementary school. Girls quickly pick up on teachers' negative messages about math—the sort that are often given out of kindness, such as: "I know this is really hard, but let's try and do it" or "I was bad at math at school" or "I never liked math." This study also highlights the link between the messages teachers give and the achievement of their students.

Wherever you are on your own mindset journey, whether these ideas are new to you or you are a mindset expert, I hope that the data and ideas I share in this book will help you and your students see mathematics—any level of mathematics—as both reachable and enjoyable. In the next chapters, Two through Eight, I will share the many strategies I have collected over years of research and practical experiences in classrooms for encouraging a growth mindset in math classrooms and homes—strategies to give students the experiences that allow them to develop strong *mathematical mindsets*.

CHAPTER 2

The Power of Mistakes and Struggle

I started teaching workshops on how to teach mathematics for a growth mindset with my graduate students from Stanford (Sarah Kate Selling, Kathy Sun, and Holly Pope) after principals of schools in California told me that their teachers had read Dweck's books and were "totally on board" with the ideas but didn't know what it meant for their mathematics teaching. The first workshop took place on Stanford's campus, in the light and airy Li Ka Shing center. For me, one of the highlights of that first workshop was when Carol Dweck met with the teachers and said something that amazed them: "Every time a student makes a mistake in math, they grow a synapse." There was an audible gasp in the room as teachers realized the significance of this statement. One reason it is so significant is that it speaks to the huge power and value of mistakes, although students everywhere think that when they make a mistake it means that they are not a "math person" or worse, that they are not smart. Many good teachers have told students for years that mistakes are useful and they show that we are learning, but the new evidence on the brain and mistakes says something much more significant.

Psychologist Jason Moser studied the neural mechanisms that operate in people's brains when they make mistakes (Moser et al., 2011). Jason and his group found something fascinating. When we make a mistake, the brain has two potential responses. The first, called an ERN response, is increased electrical activity when the brain experiences conflict between a correct response and an error. Interestingly, this brain activity occurs whether or not the person making the response knows they have made an error. The second response, called a Pe, is a brain signal reflecting conscious attention to mistakes. This happens when there is awareness that an error has been made and conscious attention is paid to the error.

When I have told teachers that mistakes cause your brain to spark and grow, they have said, "Surely this happens only if students correct their mistake and go on to solve the problem." But this is not the case. In fact, Moser's study shows us that we don't even have to be aware we have made a mistake for brain sparks to occur. When teachers ask me how this can be possible, I tell them that the best thinking we have on this now is that the brain sparks and grows when we make a mistake, even if we

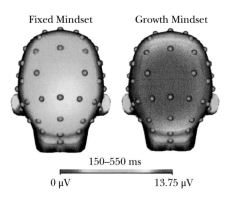

Fixed Mindset   Growth Mindset

150–550 ms

0 μV   13.75 μV

FIGURE 2.1   Brain activity in individuals with a fixed and a growth mindset
Source: Moser et al., 2011.

are not aware of it, because it is a time of struggle; the brain is challenged, and this is the time when the brain grows the most.

In Moser and his colleagues' study, the scientists looked at people's mindsets and compared mindsets with their ERN and Pe responses when they made mistakes on questions. Moser's study produced two important results. First, the researchers found that the students' brains reacted with greater ERN and Pe responses—electrical activity—when they made mistakes than when their answers were correct. Second, they found that the brain activity was greater following mistakes for individuals with a growth mindset than for individuals with a fixed mindset. Figure 2.1 represents brain activity in individuals with a fixed or growth mindset, with the growth mindset brains lighting up to a much greater extent when mistakes were made.

The fact that our brains react with increased activity when we make a mistake is hugely important. I will return to this finding in a moment.

The study also found that individuals with a growth mindset had a greater awareness of errors than individuals with a fixed mindset, so they were more likely to go back and correct errors. This study supported other studies (Mangels, Butterfield, Lamb, Good, & Dweck, 2006) showing that students with a growth mindset show enhanced brain reaction and attention to mistakes. All students responded with a brain spark—a synapse—when they made mistakes, but having a growth mindset meant that the brain was more likely to spark again, showing awareness that a mistake had been made. Whether it is mathematics, teaching, parenting, or other areas of your life, it is really important to believe in yourself, to believe that you can do anything. Those beliefs can change everything.

The recent neurological research on the brain and mistakes is hugely important for us as math teachers and parents, as it tells us that making a mistake is a very good thing. When we make mistakes, our brains spark and grow. Mistakes are not only opportunities for learning, as students consider the mistakes, but also times when our brains grow, even if we don't know we have made a mistake. The power of mistakes is critical information, as children and adults everywhere often feel terrible when they make a mistake in math. They think it means they are not a math person, because they have been brought up in a performance culture (see Boaler, 2014b) in which mistakes are not valued—or worse, they are punished. We want students to make mistakes, yet many classrooms are designed to give students work that they will get correct. Later in the book I will show the sorts of math questions that engage students and enable their brains to grow, along with the teaching and parent messages that need to accompany them.

Countries that top the world in math achievement, such as China, deal with mistakes very differently. I recently watched a math lesson in a second-grade classroom in Shanghai, the area of China where students score at the highest levels in the country and the world. The teacher gave the students deep conceptual problems to work on and then called on them for their answers. As the students happily shared their work, the interpreter leaned over and told me that the teacher was choosing students who had made mistakes. The students were proud to share their mistakes, as mistakes were valued by the teacher. In Chapter Nine I share a short and very interesting extract from one of the lessons in China.

The various research studies on mistakes and the brain not only show us the value of mistakes for everyone; they also show us that students with a growth mindset have greater brain activity related to error recognition than those with a fixed mindset. This is yet another reason why a growth mindset is so important to students as they learn mathematics as well as other subjects.

Moser's study, showing that individuals with a growth mindset have more brain activity when they make a mistake than those with a fixed mindset, tells us something else very important. It tells us that the ideas we hold about ourselves—in particular, whether we believe in ourselves or not—change the workings of our brains. If we believe that we can learn, and that mistakes are valuable, our brains grow to a greater extent when we make a mistake. This result is highly significant, telling us again how important it is that all students believe in themselves—and how important it is for all of us to believe in ourselves, particularly when we approach something challenging.

Mistakes in Life

Studies of successful and unsuccessful business people show something surprising: what separates the more successful people from the less successful people is not the number of their successes but the number of mistakes they make, with the more successful people making *more* mistakes. Starbucks is one of the world's most successful companies, and Howard Schultz, its founder, one of the most successful entrepreneurs of our time. When Schultz started what would later became Starbucks, he modeled the stores on Italian coffee shops. The United States did not have many coffee shops at the time, and Schultz had admired the coffee shops of Italy. He set up the early stores with servers wearing bow ties, which they found uncomfortable, and opera music played loudly as customers drank their coffee. The approach was not well received by American customers, and the team went back to the drawing board, making many more mistakes before eventually producing the Starbucks brand.

Peter Sims, a writer for the *New York Times*, has written widely about the importance of mistakes for creative, entrepreneurial thinking (Sims, 2011). He points out: "Imperfection is a part of any creative process and of life, yet for some reason we live in a culture that has a paralyzing fear of failure, which prevents action and hardens a rigid perfectionism. It's the single most disempowering state of mind you can have if you'd like to be more creative, inventive, or entrepreneurial."

FIGURE 2.2   Feel comfortable being wrong

FIGURE 2.3   Try seemingly wild ideas

FIGURE 2.4   Are open to different experiences

FIGURE 2.5   Play with ideas without judging them

He also summarizes the habits of successful people in general, saying that successful people:

> Feel comfortable being wrong
>
> Try seemingly wild ideas
>
> Are open to different experiences
>
> Play with ideas without judging them
>
> Are willing to go against traditional ideas
>
> Keep going through difficulties

This summer I taught a new online class for students, How to Learn Math: For Students; at the time of this writing it has been taken by over 100,000 students. The class is designed to give students a growth mindset, to show them math as engaging and exciting, and to teach them important math strategies that I will share in this book. (The course can be easily accessed at https://www.youcubed .org/category/mooc/.)

I taught the class with some of my Stanford undergraduates, who acted out the different habits that Peter Sims described, which Colin, the course producer, made more interesting with the addition of some props and characters! The undergraduates featured are Carinne Gale (Figure 2.2), Montse Cordero (Figures 2.3, 2.4, and 2.7), Devin Guillory (Figure 2.5), and Hugo Valdivia (Figure 2.6).

These different habits are just as important in math class as they are in life, but they are often startlingly absent in math class and when students work on math at home. We want students to feel free as they work on math, free to try different ideas, not fearing that they might be wrong. We want students to be open to approaching mathematics differently, being willing to play with mathematics tasks, trying "seemingly wild ideas" (see Chapter Five). We want them to go against traditional ideas—rejecting notions that some people can do math and some can't, and of course keeping going when math is hard, even when they cannot see an immediate solution.

FIGURE 2.6 Are willing to go against traditional ideas

FIGURE 2.7 Keep going through difficulties
Source: Images from *How to Learn Math: For Students*. Jo Boaler Standford Online Course. Featuring, in order: Carinne Gale, Montse Cordero, Devine Guillory, Hugo Valdivia

How Can We Change the Ways Students View Mistakes?

One of the most powerful moves a teacher or parent can make is in changing the messages they give about mistakes and wrong answers in mathematics. I recently received a very moving video from a teacher who took my online class and started the year teaching a class of failing students the importance and value of mistakes. The students completely changed over the year, picking themselves up from past failures and reengaging positively with math. The teacher sent a video of the students reflecting, in which they talk about the message that mistakes grow your brain, changing everything for them. They said that they had previously thought of themselves as failures, a mindset that had hampered their progress. Their new teacher gave them messages and teaching methods that caused them to shed their years of mathematics fear and approach the subject with new drive. When we teach students that mistakes are positive, it has an incredibly liberating effect on them.

In my online class for teachers and parents I shared the new information about mistakes and posed a challenge as one of the class activities. I asked participants to design a new activity that would reposition mistakes in classrooms or in homes. One of my favorite responses to this question came from a teacher who told me she would start the class by asking students to crumple up a piece of paper and throw it at the board with the feelings they had when they made a mistake in math. The students were invited to let out their feelings—usually ones of frustration—by hurling their crumpled paper at the board (see Figure 2.8). She then asked students to retrieve their paper, smooth it out, and trace all the crumple lines on the paper with colored markers, which represented their brain growth. The students were asked to keep the pieces of paper in their folders during the school year as a reminder of the importance of mistakes.

A few years ago I started working with Kim Halliwell, an inspirational teacher who is one of a group of teachers in Vista Unified School district with whom I have worked closely for over two years. When I visited Kim's classroom last year I saw the walls covered with lovely student drawings of brains, filled with positive messages about brain growth and mistakes (see Figure 2.9).

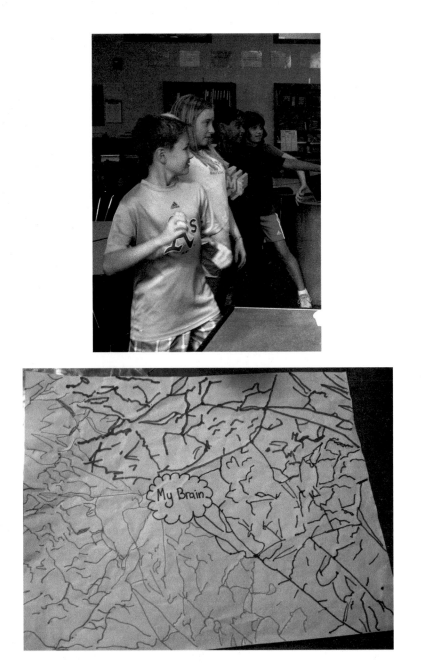

FIGURE 2.8 Students learn about brain growth

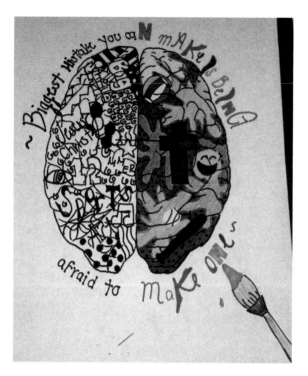

FIGURE 2.9   Student's poster of brain with messages

Kim explained to me that she had asked the students to take their favorite messages about brain growth from those they had reviewed together and put them into drawings of their brains.

Another strategy for celebrating mistakes in class is to ask students to submit work of any form—even test papers (although the less we test students the better, as I will share in Chapter Eight); teachers then highlight their "favorite mistakes." Teachers should share with students that they are looking for their favorite mistakes, which should be conceptual mistakes, not numerical errors. Teachers can then share the mistakes with the class and launch a class discussion about where the mistake comes from and why it is a mistake. This is also a good time to reinforce important messages—that when the student made this mistake, it was good, because they were in a stage of cognitive struggle and their brain was sparking and growing. It is also good to share and discuss mistakes, because if one student makes a mistake we know others are making them also, so it is really helpful for everyone to be able to think about them.

If students are graded for math work (an unhelpful practice that I will discuss later) and they are graded down for making mistakes, then they receive a very negative message about mistakes and mathematics learning. To teach students a growth mindset and general positive messages about mathematics learning, teachers should abandon testing and grading as much as possible (see Chapter Eight); if they do continue to test and grade, they should give the same grade, or higher, for mistakes, with a message attached that the mistake is a perfect opportunity for learning and brain growth.

It is important to publicly value mistakes in class, but teachers also need to give positive messages about mistakes in one-to-one interactions. My own daughter was given very damaging messages by teachers in her early years of schooling, which gave her a fixed mindset at an early age. When she was four and five she suffered from hearing difficulties (which at the time none of us knew about). Because of this, teachers decided she was not capable and gave her easy work to do. She was extremely aware of this and came home to me when she was only four asking why the other children were given harder work to do. We know that students spend a lot of time in school trying to work out what their teacher thinks of them, and she could tell that her teachers did not regard her highly. Because of this, she became convinced that she was stupid. Now, at 12, after three years in a wonderful elementary school that quickly identified her fixed mindset and saw that it was holding her back, she is a changed person and loves math.

When my daughter was in fourth grade and still suffering from a fixed mindset, she and I visited a third-grade classroom at her school. The teacher put two number problems on the whiteboard, and my daughter got one right and one wrong. When she realized she had made a mistake, she immediately reacted badly, saying she was terrible at math, and she wasn't even as good as a third grader. I took that moment to communicate something very direct and important. I said "Do you know what just happened? When you got that answer wrong your brain grew, but when you got the answer right, nothing happened in your brain; there was no brain growth." This is the sort of one-to-one interaction teachers can have with their students when they make mistakes. She looked at me with widening eyes, and I knew that she had understood the importance of the idea. Now, as she enters sixth grade, she is a different student: she embraces mistakes and feels positive about herself. This has come about not from teaching her more math or other work, but by teaching her to have a growth mindset.

In the 1930s the Swiss psychologist Jean Piaget, one of the world's leading psychologists, rejected the idea that learning was about memorizing procedures; he pointed out that true learning depends on an understanding of how ideas fit together. He proposed that students have mental models that map out the way ideas fit together, and when their mental models make sense to students, they are in a state he called equilibrium (see, for example, Piaget, 1958, 1970). When students encounter new ideas, they strive to fit the new ideas into their current mental models, but when these do not appear to fit, or their existing model needs to change, they enter a state Piaget called disequilibrium. A person in disequilibrium knows that new information cannot be incorporated into their learning models, but the new information also cannot be rejected because it makes sense, so they work to adapt their models. The process of disequilibrium sounds uncomfortable for learners, but it is disequilibrium, Piaget claims, that leads to true wisdom. Piaget showed learning as a process of moving from equilibrium, where everything fits together well, to disequilibrium, where a new idea does not fit, to a new state of equilibrium. This process, Piaget states, is essential to learning (Haack, 2011).

In Chapter Four, when I consider the act of practicing in mathematics and the forms of practice that are and are not helpful, I will show that one of the problems with our current version of mathematics education is that students are given repetitive and simple ideas that do not help them to move into the important state of disequilibrium. We know that individuals with a high tolerance for ambiguity make the transition from disequilibrium to equilibrium more readily—yet another reason we need to give students more experiences of mathematical ambiguity and risk taking. Later chapters will give ideas for ways to do this.

The research on mistakes and on disequilibrium has huge implications for mathematics classrooms, not only in the ways mistakes are handled but also in the work given to students. If we want students to be making mistakes, we need to give them challenging work that will be difficult for them, that will prompt disequilibrium. This work should be accompanied by positive messages about mistakes, messages that enable students to feel comfortable working on harder problems, making mistakes, and continuing on. This will be a big change for many teachers who currently plan the tasks given in mathematics classrooms to ensure student success and therefore give students questions that they usually answer correctly. This means that students are not being stretched enough, and they are not getting sufficient opportunities to learn and to grow their brains.

In workshops with Carol Dweck I often hear her tell parents to communicate to their children that it is not impressive to get work correct, as that shows they were not learning. Carol suggests that if children come home saying they got all their questions right in class or on a test, parents should say: "Oh, I'm sorry; that means you were not given opportunities to learn anything." This is a radical message, but we need to give students strong messages to override an idea they often get in school—that it is most important to get everything correct, and that correctness is a sign of intelligence. Both Carol and I try to reorient teachers so they value correct work less and mistakes more.

Sandie Gilliam is an incredible teacher whom I have observed teach over many years and whose students reach the highest levels and love math. One day I was observing her on the first day of class teaching high school sophomores. After students worked for a while, she noticed a student make a mistake and become aware of it. She approached the boy and asked him if he would show his mistake on the board—he looked at her with uncertainty and said, "But I got the answer wrong." Sandie replied that was why she wanted him to share his work, and that it would be really helpful. She told him that if he had made that mistake, others would have too, and it would be great for everyone if they discussed it. The boy agreed and shared his mistake with the rest of the class, showing it on the white board at the front. As the year went on, the sharing of mistakes became a common occurrence for different students. I often show a video of Sandie's students that helps teachers and policy makers see just what students can do if they are given powerful math teaching. In one of my favorite videos, we see Sandie's students work together to solve a complex problem on the board (see Figure 2.10).

The students struggle to solve the problem, and they listen to each other as different students offer ideas. They make mistakes and take wrong turns, but eventually they solve the problem, with many different students contributing. It is a powerful case of students using standard mathematical methods and mathematical practices (as recommended in the Common Core State Standards—CCSS). They combine their own thoughts and ideas with methods they know to solve an irregular applied problem of the type they will face in the world. Experienced teachers often watch the video and point out that they can see that the students feel really comfortable offering different ideas and are not afraid of being wrong. There is a reason that students are able to work at high levels unhampered by a fear of making mistakes—Sandie has taught them to embrace mistakes, and she values them in all of her teaching.

I recently worked on a research study with Carol Dweck, Greg Walton, Carissa Romero, and Dave Paunesku at Stanford; they are the team members who have delivered many important

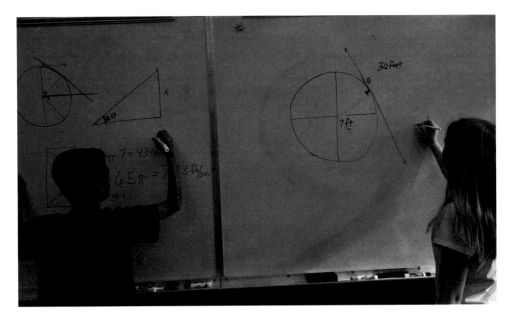

FIGURE 2.10   Solving the skateboard problem

interventions that improve students' mindsets and sense of belonging in school (for more about the Project for Education Research That Scales or PERTS, see https://www.perts.net/). In our study, we gave an intervention to math teachers, teaching them the value of mistakes and some of the teaching ideas I have shared in this chapter. We quickly found that teachers who completed the intervention had significantly more growth mindsets, had more positive feelings about mistakes in mathematics, and reported using mistake-promoting ideas in their classrooms. There are other important changes teachers can make in classrooms, and later chapters will explore these ideas; for now, one of the most important changes that a teacher (or parent) can easily make—one that has the power to make a huge difference for students—is changing the messages that students receive about mistakes. In the next chapter I will talk about the importance of changing something just as important—the mathematics. When mathematics is taught as an open and creative subject, all about connections, learning, and growth, and mistakes are encouraged, incredible things happen.

CHAPTER 3

The Creativity and Beauty in Mathematics

What is mathematics, really? And why do so many students either hate it or fear it—or both? Mathematics is *different* from other subjects, not because it is right or wrong, as many people would say, but because it is taught in ways that are not used by other subject teachers, and people hold beliefs about mathematics that they do not hold about other subjects. One way that mathematics is different is that it is often thought of as a performance subject—if you ask most students what they think their role is in math classrooms, they will tell you it is to get questions right. Students rarely think that they are in math classrooms to appreciate the beauty of mathematics, to ask deep questions, to explore the rich set of connections that make up the subject, or even to learn about the applicability of the subject; they think they are in math classrooms to perform. This was brought home to me recently when a colleague, Rachel Lambert, told me her six-year-old son had come home saying he didn't like math; when she asked him why, he said that "math was too much answer time and not enough learning time." Students from an early age realize that math is different from other subjects, and that learning gives way to answering questions and taking tests—performing.

The testing culture in the United States, which is more pervasive in math than in other subjects, is a large part of the problem. When sixth graders in my local district came home saying that they had a test on the first day of middle school, it was in one subject only: math. Most students and parents accept the testing culture of math—as one girl said to me, "Well, the teacher was just finding out what we know." But why does this happen only in math? Why do teachers not think they have to find out what students know on the first day through a test in other subjects? And why do some educators not realize that constant testing does more than test students, which has plenty of its own problems—it also makes students think that is what math is—producing short answers to narrow questions under pressure? It is no wonder that so many students decide mathematics is not for them.

There are other indications that math is different from all other subjects. When we ask students what math is, they will typically give descriptions that are very different from those given by experts in the field. Students will typically say it is a subject of calculations, procedures, or rules. But when we ask mathematicians what math is, they will say it is the study of patterns; that it is an aesthetic, creative, and beautiful subject (Devlin, 1997). Why are these descriptions so different? When we ask students of English literature what the subject is, they do not give descriptions that are markedly different from what professors of English literature would say.

Maryam Mirzakani is a mathematician at Stanford who recently won the Fields Medal, the world's top prize in mathematics. Maryam is an amazing woman who studies hyperbolic surfaces and who recently produced what has been called "the theory of the decade." In news articles on her work she is shown sketching ideas on large pieces of paper on her kitchen table, as her work is almost entirely visual. Recently I was chairing the PhD viva of one of Maryam's students. A viva is the culminating exam for PhD students when they "defend" the dissertations they have produced over a number of years in front of their committee of professors. I walked into the math department at Stanford that day, curious about the defense I was to chair. The room in which the defense was held was small, with windows overlooking Stanford's impressive Palm Drive, the entrance to the university, and it was filled with mathematicians, students, and professors who had come to watch or judge the defense. Maryam's student was a young woman names Jenya Sapir, who strode up and down that day, sharing drawings on different walls of the room, pointing to them as she made conjectures about the relationships between lines and curves on her drawings. The mathematics she described was a subject of visual images, creativity, and connections, and it was filled with uncertainty (see Figures 3.1 and 3.2).

FIGURE 3.1   Maryam Mirzakhani, winner of the Fields Medal, 2015
Source: Photo courtesy of Jan Vondrak and Maryam Mirzakhani.

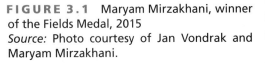

Three or four times in the defense, professors asked questions, to which the confident young woman simply answered, "I don't know." Often the professor added that she or he did not know either. It would be very unusual in a defense of an education PhD for a student to give the answer: "I don't know," and it would be frowned upon by some professors. But mathematics, real mathematics, is a subject full of uncertainty; it is about explorations, conjectures, and interpretations, not definitive answers. The professors thought it was perfectly reasonable that she did not know the answers to some of the questions, as her work was entering uncharted territories. She passed the PhD exam with flying colors.

This does not mean that there are no answers in mathematics. Many things are known and are important for students to learn. But somehow school mathematics has become so far removed from real mathematics that if I had taken most school students into the mathematics department defense that day, they would not have recognized the subject before them. This wide gulf between real mathematics and school mathematics is at the heart of the math problems we face in education. I strongly believe that if school math classrooms presented the true nature

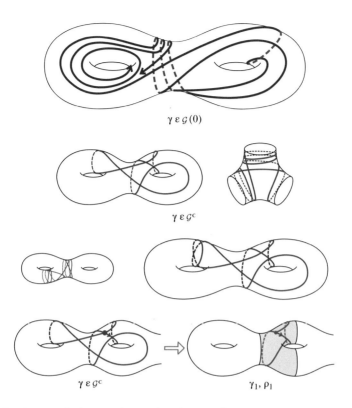

$\gamma \, \varepsilon \, \mathcal{G}(0)$

$\gamma \, \varepsilon \, \mathcal{G}^c$

$\gamma \, \varepsilon \, \mathcal{G}^c$

γ_1, ρ_1

FIGURE 3.2 Some of the mathematics discussed in Jenya Sapir's mathematics PhD defense
Source: Courtesy of Jenya Sapir.

of the discipline, we would not have this nationwide dislike of math and widespread math underachievement.

Mathematics is a cultural phenomenon; a set of ideas, connections, and relationships that we can use to make sense of the world. At its core, mathematics is about patterns. We can lay a mathematical lens upon the world, and when we do, we see patterns everywhere; and it is through our understanding of the patterns, developed through mathematical study, that new and powerful knowledge is created. Keith Devlin, a top mathematician, has dedicated a book to this idea. In his book *Mathematics: The Science of Patterns* he writes:

> As the science of abstract patterns, there is scarcely any aspect of our lives that is not affected, to a greater or lesser extent, by mathematics; for abstract patterns are the very essence of thought, of communication, of computation, of society, and of life itself. (Devlin, 1997)

Knowledge of mathematical patterns has helped people navigate oceans, chart missions to space, develop technology that powers cell phones and social networks, and create new scientific and medical knowledge, yet many school students believe that math is a dead subject, irrelevant to their futures.

To understand the real nature of mathematics it is helpful to consider the mathematics in the world—the mathematics of nature. The patterns that thread through oceans and wildlife, structures and rainfall, animal behavior, and social networks have fascinated mathematicians for centuries. Fibonacci's pattern is probably the best known of all patterns. Fibonacci was an Italian mathematician who published a pattern in Italy in 1202 that became known as the Fibonacci sequence. Now it is known that the pattern appeared centuries earlier, as early as 200 B.C. in India. This is Fibonacci's famous sequence:

$$1, 1, 2, 3, 5, 8, 13, 21, 34, 55 \ldots$$

The first two numbers are 1 and 1; the rest of the numbers can be derived from adding the previous two numbers. There is something really interesting about Fibonacci's pattern. If we move along the sequence dividing each number by the one before it, we derive a ratio that gets closer and closer to 1.618:1. This is known as the *golden ratio* and this ratio exists throughout nature. The spirals in pine cones, flowers, and pineapples, for example, all produce the golden ratio.

If we consider snowflakes, we see something else that is interesting. Each snowflake is unique, but they are all unified by a single pattern. Snowflakes all follow the general shape of a hexagon, so they almost always have six points (see Figures 3.3 and 3.4). There is a reason for this: snowflakes are made up of water molecules, and when water freezes, it does so in a pattern of repeating hexagons.

Mathematics is also used by animals. When I taught an online class for students of mathematics recently, taken by over 100,000 students, I showed them the mathematics used by animals, which the students found really interesting. Dolphins, for example, emit a sound to help them find each other in the water (see Figure 3.5).

Dolphins make their characteristic clicking sounds, which rebound off other objects and echo back to the dolphin. The dolphin then uses the length of time it takes for the sound to come back and the quality of the sound to know where their friends are. Intuitively they are calculating a rate—the same rate question that students are given in algebra class (often over and over again, with no connection to a real situation). I joked with my online students that if dolphins could speak human they could become algebra teachers!

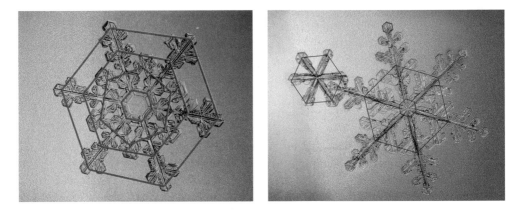

FIGURE 3.3   The mathematics in snowflakes

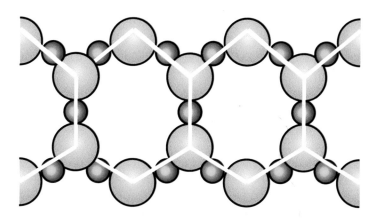

FIGURE 3.4   Water molecules

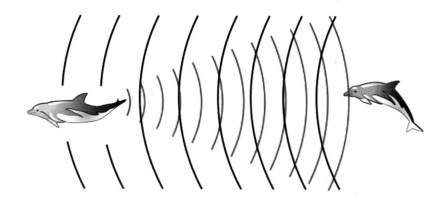

FIGURE 3.5   Dolphins communicate

In conducting research for my online class, my student Michaela found that spiders are experts in spirals. When a spider makes a web, it first creates a star shape between two sturdy vertical supports, such as tree branches. Then the spider begins to make a spiral. It needs to make this spiral as quickly as possible to strengthen the star, so the spider chooses to make a logarithmic spiral. In logarithmic spirals, the distances between each successive turn around the center increase by the same factor every time (see Figure 3.6).

This means that the spiral expands outward more and more quickly the larger it gets. But this logarithmic spiral leaves a lot of space in the web, so the spider starts a second, denser spiral. This new spiral is an *arithmetic* spiral, which means that the distances between the turns of the spiral are all the same. This second spiral takes the spider much longer to make because it has to take many more trips around the center of the star, but it helps the spider catch more insects because it eliminates large spaces from the web. This amazing feat of engineering

FIGURE 3.6   A spider web

could be constructed using calculations, but the spider intuitively uses mathematics in creating and using its own algorithm. For more examples of the mathematics used by animals, see Devlin (2006).

When I showed these ideas to students in my online class, a number of them were resistant, saying that the mathematics in nature and used by animals was not math. The students recognized only a very different math—a math of numbers and calculations. My goal was to push the students to see math broadly and to open their eyes to real mathematics, and the course was very successful in doing this. At the end of the course students were given a survey, and 70% of them said the course had changed their view of what mathematics is. Importantly, 75% had a stronger belief that they could do well in math.

Mathematics exists throughout nature, art, and the world, yet most school students have not heard of the golden ratio and do not see mathematics as the study of patterns. When we do not show the breadth of mathematics to students, we deny them the chance to experience the wonder of mathematics.

I am not the only person who has argued that school mathematics is not real mathematics. Reuben Hersh, a mathematician, wrote a fascinating book entitled *What Is Mathematics, Really?* (1999) in which he argues that mathematics is severely misrepresented in classrooms. Most students think of mathematics as a series of answers—answers to questions that nobody has asked. But Hersh points out that it is

> the questions that drive mathematics. Solving problems and making up new ones is the essence of mathematical life. If mathematics is conceived apart from mathematical life, of course it seems—dead.

Numerous research studies (Silver, 1994) have shown that when students are given opportunities to pose mathematics problems, to consider a situation and think of a mathematics question to ask of it—which is the essence of real mathematics—they become more deeply engaged and perform at higher levels. But this rarely happens in mathematics classrooms. In *A Beautiful Mind*, the box office movie hit, viewers watch John Nash (played by Russell Crowe) strive to find an interesting question to ask—the critical and first stage of mathematical work. In classrooms students do not experience this important mathematical step; instead, they spend their time answering questions that seem dead to them, questions they have not asked.

In my book *What's Math Got to Do with It?* I describe a classroom approach that was based on the posing of mathematics questions (Boaler, 2015a). The teacher, Nick Fiori, gave his students mathematical situations such as pine cones, SET playing cards, colored beads, dice, and nuts and bolts and asked the students to pose their own questions. This was a hard adjustment for students at first, but gradually and excitedly they learned to use their own ideas, conduct mathematical inquiries, and learn new methods when there was a purpose in doing so.

Over the years, school mathematics has become more and more disconnected from the mathematics that mathematicians use and the mathematics of life. Students spend thousands of hours in classrooms learning sets of procedures and rules that they will never use, in their lives or in their work. Conrad Wolfram is a director of Wolfram-Alpha, one of the most important mathematical companies in the world. He is also an outspoken critic of traditional mathematics teaching, and he argues strongly that mathematics does not equal calculating. In a TED talk watched by over a million people, Wolfram (2010) proposes that working on mathematics has four stages:

1. Posing a question

2. Going from the real world to a mathematical model

3. Performing a calculation

4. Going from the model back to the real world, to see if the original question was answered

The first stage involves asking a good question of some data or a situation—the first mathematical act that is needed in the workplace. The fastest-growing job in the United States is that of data analyst—someone who looks at the "big data" that all companies now have and asks important questions of the data. The second stage Wolfram describes is setting up a model to answer the question; the third is performing a calculation, and the fourth is turning the model back to the world to see whether the question is answered. Wolfram points out that 80% of school mathematics is spent on stage 3—performing a calculation by hand—when that is the one stage that employers do not need workers to be able to do, as it is performed by a calculator or computer. Instead, Wolfram proposes that we have students working on stages 1, 2, and 4 for much more of their time in mathematics classes.

What employers need, he argues, is people who can ask good questions, set up models, analyze results, and interpret mathematical answers. It used to be that employers needed people to calculate; they no longer need this. What they need is people to think and reason.

The Fortune 500 comprises the top 500 companies in the United States. Forty-five years ago, when companies were asked what they most valued in new employees, the list looked like this:

TABLE 3.1 Fortune 500 "most valued" skills in 1970

1	Writing
2	Computational Skills
3	Reading Skills
4	Oral Communications
5	Listening Skills
6	Personal Career Development
7	Creative Thinking
8	Leadership
9	Goal Setting/Motivation
10	Teamwork
11	Organizational Effectiveness
12	Problem Solving
13	Interpersonal Skills

Computation was second in the list. In 1999 the list had changed, as Table 3.2 shows.

Computation has dropped to the second-from-the-last position, and the top places have been taken by teamwork and problem solving.

Parents often do not see the need for something that is at the heart of mathematics: the discipline. Many parents have asked me: What is the point of my child explaining their work if they can get the answer right? My answer is always the same: Explaining your work is what, in mathematics, we call reasoning, and reasoning is central to the discipline of mathematics. Scientists prove or disprove theories by producing more cases that do or do not work, but mathematicians prove theories through mathematical reasoning. They need to produce arguments that convince other mathematicians by carefully reasoning their way from one idea to another, using logical connections. Mathematics is a very social subject, as proof comes about when mathematicians can convince other mathematicians of logical connections.

A lot of mathematics is produced through collaborations between mathematicians; Leone Burton studied the work of mathematicians and found that over half of their publications were produced collaboratively (Burton, 1999). Yet many mathematics classrooms are places where

TABLE 3.2   Fortune 500 "most valued" skills in 1999

1	Teamwork
2	Problem Solving
3	Interpersonal Skills
4	Oral Communications
5	Listening Skills
6	Personal Career Development
7	Creative Thinking
8	Leadership
9	Goal Setting/Motivation
10	Writing
11	Organizational Effectiveness
12	Computational Skills
13	Reading Skills

students complete worksheets in silence. Group and whole class discussions are really important. Not only are they the greatest aid to understanding—as students rarely understand ideas without talking through them—and not only do they enliven the subject and engage students, but they teach students to reason and to critique each other's reasoning, both of which are central in today's high-tech workplaces. Almost all new jobs in today's technological world involve working with massive data sets, asking questions of the data and reasoning about pathways. Conrad Wolfram told me that anyone who cannot reason about mathematics is ineffective in today's workplace. When employees reason and talk about mathematical pathways, other people can develop new ideas based on the pathways as well as see if a mistake has been made. The teamwork that employers value so highly is based upon mathematical reasoning. People who just give answers to calculations are not useful in the workplace; they must be able to reason through them.

We also want students reasoning in mathematics classrooms because the act of reasoning through a problem and considering another person's reasoning is *interesting* for students. Students and adults are much more engaged when they are given open math problems and allowed to come up with methods and pathways than if they are working on problems that require a calculation and answer. I will be showing many of the good, rich mathematics problems that require reasoning, and explain some ways to design them, in Chapter Five.

Another serious problem we face in math education is that people believe that mathematics is all about calculating and that the best mathematics thinkers are those who calculate the fastest. Some people believe something even worse—that you have to be *fast* at math to be *good* at math. There are strong beliefs in society that if you can do a calculation quickly then you are a true math person and that you are "smart." Yet mathematicians, whom we could think of as the most capable math people, are often slow with math. I work with many mathematicians, and they are simply not fast math thinkers. I don't say this to be disrespectful to mathematicians; they are slow because they think carefully and deeply about mathematics.

Laurent Schwartz won the Fields Medal in mathematics and was one of the greatest mathematicians of his time. But when he was in school, he was one of the slowest math thinkers in his class. In his autobiography, *A Mathematician Grappling with His Century* (2001), Schwartz reflects on his school days and how he felt "stupid" because his school valued fast thinking, but he thought slowly and deeply:

> I was always deeply uncertain about my own intellectual capacity; I thought I was unintelligent. And it is true that I was, and still am, rather slow. I need time to seize things because I always need to understand them fully. Towards the end of the eleventh grade, I secretly thought of myself as stupid. I worried about this for a long time.
>
> I'm still just as slow . . . . At the end of the eleventh grade, I took the measure of the situation, and came to the conclusion that rapidity doesn't have a precise relation to intelligence. What is important is to deeply understand things and their relations to each other. This is where intelligence lies. The fact of being quick or slow isn't really relevant. (Schwartz, 2001)

Schwartz writes, as have many other mathematicians, about the misrepresentation of mathematics in classrooms, and about mathematics being about connections and deep thinking, not fast calculation. There are many students in math classrooms who think slowly and deeply, like Laurent Schwartz, who are made to believe that they cannot be math people. Indeed, the idea that math is about fast calculations puts off large numbers of math students, especially girls, as I will talk about more in Chapters Four and Seven. Yet mathematics continues to be presented as a speed race, more than any other subject—timed math tests, flash cards, math apps against the clock. It is no wonder that students who think slowly and deeply are put off mathematics. National leaders, such as Cathy Seeley, former president of NCTM, are also working to dispel the idea that mathematics is only for fast students, offering instead a new way for teachers and students to work productively and with depth (see Seeley, 2009, 2014). The widespread myth that math is about speed is one that is very important to dispel, if we are to stop making slow and deep thinkers, such as Laurent Schwartz, and many girls (Boaler, 2002b), think that mathematics is not for them. In the next chapter I will show how mathematics—particularly numbers and calculating—can be taught in a way that values depth and not speed, that enhances brain connections and that engages many more students.

Conclusion

I started this chapter by talking about mathematics being different from other subjects. But the difference in mathematics is not because of the nature of the subject, as many people believe; rather, it is due to some serious and widespread misconceptions about the subject: that math is a subject of rules and procedures, that being good at math means being fast at math, that math is all about certainty and right and wrong answers, and that math is all about numbers. These misconceptions are held by teachers, students, and parents, and they are part of the reason that traditional, faulty, and ineffective teaching have been allowed to continue. Many parents hated mathematics in school, but they still argue for traditional teaching because they think it just has to be that way—that the unpleasant teaching they experienced is due to the unpleasant nature of mathematics. Many elementary school teachers have had terrible mathematics experiences themselves, and they struggle to teach the subject, as they think it just has to be taught as a dry set of procedures. When I show them that real mathematics is different, and that they do not have to subject students to the mathematics they experienced, they feel truly liberated and often euphoric, as Chapter Five will show. When we consider how many of the misconceptions that I have presented in this chapter are represented in math classrooms, we can more easily understand both the extent of mathematics failure in the United States and beyond and, even more important, that mathematics failure and anxiety are completely unnecessary.

When we look at mathematics in the world and the mathematics used by mathematicians, we see a creative, visual, connected, and living subject. Yet school students often see mathematics as a dead subject—hundreds of methods and procedures to memorize that they will never use, hundreds of answers to questions that they have never asked. When people are asked about how mathematics is used in the world, they usually think of numbers and calculations—of working out mortgages or sale prices—but mathematical thinking is so much more. Mathematics is at the center of thinking about how to spend the day, how many events and jobs can fit into the day, what size of space can be used to fit equipment or turn a car around, how likely events are to happen, knowing how tweets are amplified and how many people they reach. The world respects people who can calculate quickly, but the fact is, some people can be very fast with numbers and not be able to do great things with them, and others, who are very slow and make many mistakes, go on to do something amazing with mathematics. The powerful thinkers in today's world are not those who can calculate fast, as used to be true; fast calculations are now fully automated, routine, and uninspiring. The powerful thinkers are those who make connections, think logically, and use space, data, and numbers creatively.

The fact that a narrow and impoverished version of mathematics is taught in many school class-rooms cannot be blamed on teachers. Teachers are usually given long lists of content to teach, with hundreds of content descriptions and no time to go into depth on any ideas. When teachers are given lists of content to teach, they see a subject that has been stripped down to its bare parts like a dismantled bike—a collection of nuts and bolts that students are meant to shine and polish all year. Lists of contents don't include connections; they present mathematics as though connections do not even exist. I don't want students polishing disconnected bike parts all day! I want them to get onto the assembled bikes and ride freely, experiencing the pleasure of math, the joy of making connections, the euphoria of real mathematical thinking. When we open up mathematics and teach the broad, visual, creative math that I will show in this book, then we also teach math as a learning subject. It is very hard for students to develop a growth mindset if they only ever answer questions that they get either right or wrong. Such questions themselves transmit fixed messages about math. When we teach mathematics—real mathematics, a subject of depth and connections—the opportunities for a growth mindset increase, the opportunities for learning increase, and classrooms become filled with happy, excited, and engaged students. The next five chapters will be filled with ideas for ways to get this to happen, as well as the research evidence supporting them.

CHAPTER 4

Creating Mathematical Mindsets: The Importance of Flexibility with Numbers

Babies and infants love mathematics. Give babies a set of blocks, and they will build and order them, fascinated by the ways the edges line up. Children will look up at the sky and be delighted by the V formations in which birds fly. Count a set of objects with a young child, move the objects and count them again and they will be enchanted by the fact they still have the same number. Ask children to make patterns in colored blocks and they will work happily making repeating patterns—one of the most mathematical of all acts. Keith Devlin has written a range of books showing strong evidence that we are all natural mathematics users, and thinkers (see, for example, Devlin, 2006). We want to see patterns in the world and to understand the rhythms of the universe. But the joy and fascination young children experience with mathematics are quickly replaced by dread and dislike when they start school mathematics and are introduced to a dry set of methods they think they just have to accept and remember.

In Finland, one of the highest-scoring countries in the world on PISA tests, students do not learn formal mathematics methods until they are seven. In the United States, United Kingdom, and a few other countries, students start much earlier, and by the time our students are seven they have already been introduced to algorithms for adding, subtracting, multiplying, and dividing numbers and been made to memorize multiplication facts. For many students their first experience of math is one of confusion, as the methods do not make sense to them. The inquisitiveness of our children's early years fades away and is replaced by a strong belief that math is all about following instructions and rules.

The best and most important start we can give our students is to encourage them to play with numbers and shapes, thinking about what patterns and ideas they can see. In my previous book I shared the story of Sarah Flannery, who won the Young Scientist of the Year Award for inventing a new mathematical algorithm. In her autobiography she talks about the way she developed her mathematical thinking from working on puzzles in the home with her dad, and how these puzzles were more important to her than all of her years of math class (Flannery, 2002). Successful math users have an approach to math, as well as mathematical understanding, that sets them apart from less successful users. They approach math with the desire to understand it and to think about it, and with the confidence that they can make sense of it. Successful math users search for patterns and relationships and think about connections. They approach math with a *mathematical mindset,* knowing that math is a subject of growth and their role is to learn and think about new ideas. We need to instill this *mathematical mindset* in students from their first experiences of math.

Research has shown definitively the importance of a growth mindset—the belief that intelligence grows and the more you learn, the smarter you get. But to erase math failure we need students to have growth beliefs about themselves and accompany them with growth beliefs about the nature of mathematics and their role in relation to it. Children need to see math as a conceptual, growth subject that they should think about and make sense of. When students see math as a series of short questions, they cannot see the role for their own inner growth and learning. They think that math is a fixed set of methods that either they get or they don't. When students see math as a broad landscape of unexplored puzzles in which they can wander around, asking questions and thinking about relationships, they understand that their role is thinking, sense making, and growing. When students see mathematics as a set of ideas and relationships and their role as one of thinking about the ideas, and making sense of them, they have a mathematical mindset.

Sebastian Thrun, CEO of Udacity and a research professor at Stanford University, has a mathematical mindset. I started working with Sebastian two years ago. Initially I knew him as a computer science professor and the person who invented self-driving cars, taught the first MOOC, and led teams developing Google Glass and Google Maps. Sebastian moved from his highly successful online course, taken by 160,000 people, to forming Udacity, an online learning company. I started working with Sebastian when he asked for my advice on Udacity's courses. Sebastian is a very high-level user of mathematics whose many accomplishments are known around the world. He has written mathematics books that are so complex that they would, as he says, "make your head smoke." What is less well known about Sebastian is that he is highly reflective about the ways he knows and learns mathematics. When I interviewed Sebastian for my online course (How to Learn Math) for teachers and parents, he talked about the important role played by intuition in mathematics learning and problem solving, and of making sense of situations. He gave a specific example of a time when he was developing robots to be used at the Smithsonian Institution and a problem came up. The children and other visitors at the Smithsonian were creating background noise that was confusing the robots. Sebastian said that he and his team had to go back to their drawing boards and construct new mathematical pathways that would solve the problem and allow the robots to function. He eventually solved the problem by using intuition. Sebastian describes the process of working out a mathematical solution that made sense to him intuitively, then going back and proving it using mathematical methods. Sebastian speaks strongly about never going forward in mathematics unless something makes intuitive sense. In my online course he gives advice to children learning mathematics to never work with formulae or methods unless they make sense and to "just stop" if the methods don't make sense.

So how do we develop mathematical mindsets in students so that they are willing to approach math with sense making and intuition? Before they start school, the task is straightforward. It means asking children to play with puzzles, shapes, and numbers and think about their relationships. But in the early years of school we live in a system whereby students are required, from an early age, to learn many formal mathematical methods, such as those used to add, subtract, divide, and multiply numbers. This is the time when students stray from mathematical mindsets and develop fixed, procedural mindsets. This is the time when it is most critical that teachers and parents introduce mathematics as a flexible conceptual subject that is all about thinking and sense making. The domain of early number work gives us the perfect example of the two mindsets that can develop in students, one that is negative and leads to failure and one that is positive and leads to success.

Number Sense

Eddie Gray and David Tall are two British researchers who worked with students, aged 7 to 13, who had been nominated by their teachers as being either low-, middle-, or high-achieving students (Gray & Tall, 1994). All of the students were given number problems, such as adding or subtracting two numbers. The researchers found an important difference between the low- and high-achieving students. The high-achieving students solved the questions by using what is known as number sense—they interacted with the numbers flexibly and conceptually. The low-achieving students used no number sense and seemed to believe that their role was to recall and use a standard method even when this was difficult to do. For example, when students were given a problem such as 21 – 6, the high-achieving students made the problem easier by changing it to 20 – 5, but the low-achieving students counted backward, starting at 21 and counting down, which is difficult to do and prone to error. After extensive study of the different strategies that the students used, the researchers concluded that the difference between high- and low-achieving students was not that the low-achieving students knew less mathematics, but that they were interacting with mathematics differently. Instead of approaching numbers with flexibility and number sense, they seemed to cling to formal procedures they had learned, using them very precisely, not abandoning them even when it made sense to do so. The low achievers did not *know less,* they just did not use numbers flexibly—probably because they had been set on the wrong pathway, from an early age, of trying to memorize methods and number facts instead of interacting with numbers flexibly (Boaler, 2015). The researchers pointed out something else important—the mathematics the low achievers were using was a harder mathematics. It is much easier to subtract 5 from 20 than to start at 21 and count down 16 numbers. Unfortunately for low achievers, they are often identified as struggling with math and therefore given more drill and practice—cementing their beliefs that math success means memorizing methods, not understanding and making sense of situations. They are sent down a damaging pathway that makes them cling to formal procedures, and as a result they often face a lifetime of difficulty with mathematics.

A mathematical mindset reflects an active approach to mathematics knowledge, in which students see their role as understanding and sense making. Number sense reflects a deep understanding of mathematics, but it comes about through a mathematical mindset that is focused on making sense of numbers and quantities. It is useful to think about the ways number sense is developed in students, not only because number sense is the foundation for all higher level mathematics (Feikes & Schwingendorf, 2008) but also because number sense and mathematical mindsets develop together, and learning about ways to develop one helps the development of the other.

Mathematics is a conceptual domain. It is not, as many people think, a list of facts and methods to be remembered.

In Figure 4.1, the purple arrows represent methods to be learned; the pink boxes represent the concepts being learned. Starting at the bottom left of the diagram, we see the method of counting. When students learn to count, they remember order and names for numbers, but they also develop the *concept* of number; that is, the idea of a number. In the early stages of learning to add numbers, students learn a method called "counting on." Counting on is used when you have two sets of numbers—for example, 15 plus 4—and you learn to count the first set: counting to 15, then continuing counting: 16–17–18–19. When students learn the method of counting on, they develop the concept of "sum." This is not a method of addition; it is a conceptual idea. In the next stage of their mathematics work, students may learn to add groups of numbers, such as three groups of 4, and as they learn to add groups, they develop the concept of a product. Again, this is not a method (of multiplication); it is a conceptual idea. The ideas of a number,

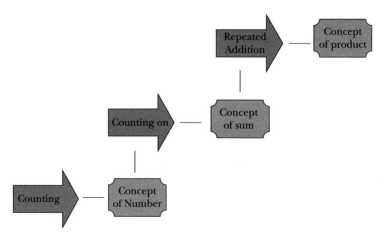

FIGURE 4.1   Mathematics methods and concepts
Source: Gray & Tall, 1994.

a sum, and a product are concepts in mathematics that students need to think deeply about. Students should learn methods, such as adding and multiplying, not as ends in themselves but as part of a conceptual understanding of numbers, sums, and products and how they relate to each other.

We know that when we learn mathematics we engage in a brain process called *compression*. When you learn a new area of mathematics that you know nothing about, it takes up a large space in your brain, as you need to think hard about how it works and how the ideas relate to other ideas. But the mathematics you have learned before and know well, such as addition, takes up a small, compact space in your brain. You can use it easily without thinking about it. The process of compression happens because the brain is a highly complex organ with many things to control, and it can focus on only a few uncompressed ideas at any one time. Ideas that are known well are compressed and filed away. William Thurston, a top mathematician who won the Fields Medal, describes compression like this:

> Mathematics is amazingly compressible: you may struggle a long time, step by step, to work through the same process or idea from several approaches. But once you really understand it and have the mental perspective to see it as a whole, there is often a tremendous mental compression. You can file it away, recall it quickly and completely when you need it, and use it as just one step in some other mental process. The insight that goes with this compression is one of the real joys of mathematics. (Thurston, 1990)

Many students do not describe mathematics as a "real joy"—in part because they are not engaging in compression. Notably, the brain can only compress concepts; it cannot compress rules and methods. Therefore students who do not engage in conceptual thinking and instead approach mathematics as a list of rules to remember are not engaging in the critical process of compression, so their brain is unable to organize and file away ideas; instead, it struggles to hold onto long lists of methods and rules. This is why it is so important to help students approach mathematics conceptually at all times. Approaching mathematics conceptually is the essence of what I describe as a mathematical mindset.

What about Math Facts?

Many people believe that it is not possible to think conceptually about mathematics all of the time because there are lots of math facts (such as $8 \times 4 = 32$) that have to be memorized. There are some math facts that it is good to memorize, but students can learn math facts and commit them to memory through conceptual engagement with math. Unfortunately, most teachers and parents think that because some areas of mathematics are factual, such as number facts, they need to be learned through mindless practice and speed drills. It is this approach to early learning about numbers that causes damage to students, makes them think that being successful at math is about recalling facts at speed, and pushes them onto a procedural pathway that works against their development of a mathematical mindset.

Math facts by themselves are a small part of mathematics, and they are best learned through the use of numbers in different ways and situations. Unfortunately, many classrooms focus on math facts in isolation, giving students the impression that math facts are the essence of mathematics, and, even worse, that mastering the fast recall of math facts is what it means to be a strong mathematics student. Both of these ideas are wrong, and it is critical that we remove them from classrooms, as they play a key role in creating math-anxious and disaffected students.

I grew up in a progressive era in England, when primary schools focused on the "whole child," and I was not presented with tables of addition, subtraction, or multiplication facts to memorize in school. I have never committed math facts to memory, although I can quickly produce any math fact, as I have number sense and I have learned good ways to think about number combinations. My lack of memorization has never held me back at any time or place in my life, even though I am a mathematics professor, because I have number sense, which is much more important for students to learn and includes learning of math facts along with deep understanding of numbers and the ways they relate to each other.

For about one-third of students, the onset of timed testing is the beginning of math anxiety (Boaler, 2014c). Sian Beilock and her colleagues have studied people's brains through MRI imaging and found that math facts are held in the working memory section of the brain. But when students are stressed, such as when they are taking math questions under time pressure, the working memory becomes blocked, and students cannot access math facts they know (Beilock, 2011). As students realize they cannot perform well on timed tests, they start to develop anxiety, and their mathematical confidence erodes. The blocking of the working memory and associated anxiety is particularly common among higher-achieving students and girls. Conservative estimates suggest that at least a third of students experience extreme stress related to timed tests, and these are not students from any particular achievement group or economic background. When we put students through this anxiety-provoking experience, we lose students from mathematics.

Math anxiety has now been recorded in students as young as five, and timed tests are a major cause of this debilitating, often lifelong condition. In my classes at Stanford University, I encounter many undergraduates who have been math traumatized, even though they are among the highest-achieving students in the country. When I ask them what led to their math aversion, many talk about timed tests in second or third grade as a major turning point when they decided that math was not for them. Some of the students, especially women, talk about the need to understand deeply (a very worthwhile goal) and being made to feel that deep understanding was not valued or offered when timed tests became a part of math class. They may have been doing other, more valuable work in their mathematics classes, focusing on sense making and understanding, but timed tests evoke such strong emotions that students can come to believe that being fast with math facts is the essence of mathematics. This is extremely unfortunate. We see the outcome of the misguided school emphasis on memorization and testing in the numbers of students dropping out of mathematics and the math crisis we currently face (see www.youcubed.org). When my own daughter started times table memorization and testing at age five, she started to come home and cry about math. This is not the emotion we want students to associate with mathematics, but as long as we keep putting students under pressure to recall facts at speed we will not erase the widespread anxiety and dislike of mathematics that pervades the United States and United Kingdom (Silva & White, 2013).

So what do we do to help students learn math facts if we do not use timed tests? The very best way to encourage the learning of facts and the development of a mathematical mindset is to offer conceptual mathematical activities that help students learn and understand numbers and number facts. Brain researchers studied students learning math facts in two ways. One approach was through strategies; for example, learning 17×8 by working out 17×10 (170) and subtracting 17×2 (34); the other strategy was memorization of the facts ($17 \times 8 = 136$). They found that the two approaches (strategies or memorization) involve two distinct pathways in the brain and that both pathways are perfectly good for lifelong use. Importantly, though, the study also found that those who learned through strategies achieved "superior performance" over those who memorized; they solved test questions at the same speed and showed better transfer to new problems. The brain researchers concluded that automaticity should be reached through understanding of numerical relations, achieved through thinking about number strategies (Delazer et al., 2005).

In another important study, researchers found that the most powerful learning occurs when we use different pathways in the brain (Park & Brannon, 2013). The left side of the brain handles factual and technical information; the right side brain handles visual and spatial information. Researchers have found that mathematics learning and performance are optimized when the two sides of the brain are communicating (Park & Brannon, 2013). Researchers also found that when students were working on arithmetic problems, such as subtraction, the highest-level performers were those who exhibited the strongest connections between the two sides of the brain. The implications of this finding are extremely important for mathematics learning, as they tell us that learning the formal abstract mathematics that makes up a lot of the school curriculum is enhanced when students are using visual and intuitive mathematical thinking.

In Youcubed's paper "Fluency without Fear," which became the focus of several major news studies, we included this evidence and activities that teachers and parents can use to enable the important brain connections. One of the math games we included in the paper became hugely popular after it was released and was tweeted around the world.

The game is played by partners. Each child has a blank 100 grid. The first partner rolls two number dice. The numbers that come up are the numbers the child uses to make an array on their 100 grid. They can put the array anywhere on the grid, but the goal is to fill up the grid to get it as full as possible. After the player draws the array on their grid, she writes in the number sentence that describes the grid. The game ends when both players have rolled the dice and cannot put any more arrays on the grid (see Figure 4.2).

In this game the students are learning number facts, such as 2×12, but they are also doing something much more important. They are thinking about the meaning of the number facts and what 2×12 represents, visually and spatially.

Another game that encourages the same powerful brain connections takes the idea of math cards, which are often used in damaging ways such as drill and speed "flash cards," and uses them very differently. The aim of the game is to match cards with the same answer, shown through different representations, with no time pressure. Teachers lay all the cards down on a table and ask students to take turns picking them; they pick as many as they find with the same answer (shown through any representation). For example, 9 and 4 can be shown with an area model, sets of objects such as dominoes, and a number sentence. When student match the cards, they should explain how they know that the different cards are equivalent. This activity again focuses

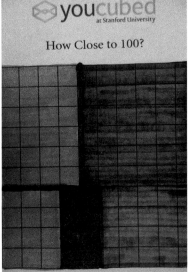

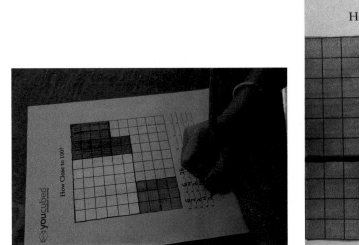

a. How close to 100   b. How close to 100-colored squares

FIGURE 4.2   How Close to 100?

on understanding multiplication, visually and spatially, encouraging brain connections at the same time as rehearsing math facts. The game can also be played with the cards face down as a memory game to add an extra challenge. A full set of cards is available at http://www.youcubed .org/wp-content/uploads/2015/03/FluencyWithoutFear-2015.pdf (see Figure 4.3).

These activities teach number sense and a mathematical mindset and encourage new brain pathways across the two brain hemispheres. The antithesis of this approach is a focus on rote memorization and speed. The more we emphasize memorization to students, the less willing they become to think about numbers and their relations and to use and develop number sense (Boaler, 2015). Some students are not as good at memorizing math facts as others. That is something to be celebrated; it is part of the wonderful diversity of life and people. Imagine how awful it would be if teachers gave tests of math facts and everyone answered them in the same way and at the same speed as though they were all robots. In a recent brain study, scientists examined students' brains as they were taught to memorize math facts. They saw that some students memorized them much more easily than others. This will come as no surprise to readers, and many of us would probably assume that those who memorized better were higher-achieving or "more intelligent" students. But the researchers found that the students who memorized more easily were not higher achieving; they did not have what the researchers described as more "math ability," nor did they have higher IQ scores (Supekar et al., 2013). The only differences the researchers found were in a brain region called the hippocampus, the area of the brain responsible for memorized facts. The hippocampus, like other brain regions, is not fixed and can grow at any time, as illustrated by the London Black Cab studies (Woollett & Maguire, 2011), but it will always be the case that some students are faster or slower when memorizing, and this *has nothing to do with mathematics potential*.

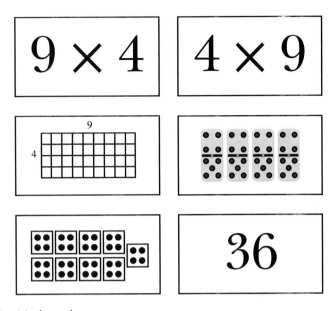

FIGURE 4.3 Math cards
Source: www.youcubed.org.

A TEACHER'S TALE OF MATH MEMORIZATION TRAUMA

In a recent professional development workshop I was conducting with teachers in California, I shared the fact that I did not memorize my times tables as a child growing up. I also shared that this has never held me back in any way, or at any time, despite engaging with and working on mathematics on a daily basis. When I told this to the room full of teachers, four of them cried. At lunch one of them explained to me, through sobs, that my making this statement had changed everything for her. As a young child she had difficulty memorizing the times tables, and her father gave her the idea that she was deficient in some way. All of her life she had felt that there was something wrong with her. She told me that her school principal was with her at the session, and she had feared that her "deficiency" would be exposed. The number of people who have been damaged by the emphasis on timed tests and the memorization of math facts in school classrooms is frighteningly large.

In order to learn to be a good English student, to read and understand novels, or poetry, students need to have memorized the meanings of many words. But no English student would say or think that learning English is about the fast memorization and recall of words. This is because we learn words by using them in many different situations—talking, reading, and writing. English teachers do not give students hundreds of words to memorize and then test them under timed conditions. All subjects require the memorization of some facts, but mathematics is the only subject in which teachers believe they should be tested under timed conditions. Why do we treat mathematics in this way? We have the research evidence that shows students can learn math facts much more powerfully with engaging activities; now is the time to use this evidence and liberate students from mathematics fear.

How Important Is Math Practice?

When I show parents and teachers the evidence that students need to engage in mathematics conceptually and visually, some parents ask, "But don't students need a lot of math practice?"—by which they mean pages of math questions given in isolation. The question of whether or how much practice students need in mathematics is an interesting one. We know that when learning happens a synapse fires, and in order for structural brain change to happen we need to revisit ideas and learn them deeply. But what does that mean? It is important to revisit mathematical ideas, but the "practice" of methods over and over again is unhelpful. When you learn a new idea in mathematics, it is helpful to reinforce that idea, and the best way to do this is by using it in different ways. We do students a great disservice when we pull out the most simple version of an idea and give students 40 questions that repeat it. Worksheets that repeat the same idea over and over turn students away from math, are unnecessary, and do not prepare them to use the idea in different situations.

In Malcolm Gladwell's bestselling book *Outliers,* he develops the idea that it takes roughly 10,000 hours of practice to achieve mastery in a field (Gladwell, 2011). Gladwell describes the achievements of famous musicians, chess players, and sports stars, and he shows something important. Many people believe that people such as Beethoven are natural geniuses, but Gladwell shows that they work hard and long to achieve their great accomplishments, with a growth mindset that supports their work. Unfortunately, I have spoken to a number of people who have interpreted Gladwell's idea to mean that students can develop expertise in math after 10,000 hours of mindless practice. This is incorrect. Expertise in mathematics requires 10,000 hours of working mathematically. We do not need students to take a single method and practice it over and over again. That is not mathematics; it does not give students the knowledge of ideas, concepts, and relationships that make up expert mathematics performance. Someone working for 10,000 hours would need to be working on mathematics as a whole, considering mathematical ideas and connections, solving problems, reasoning, and connecting methods.

Most textbook authors in the United States base their whole approach on the idea of isolating methods, reducing them to their simplest form and practicing them. This is problematic for many reasons. First, practicing isolating methods induces boredom in students; many students simply turn off when they think their role is to passively accept a method (Boaler & Greeno, 2000) and repeat it over and over again. Second, most practice examples give the most simplified and disconnected version of the method to be practiced, giving students no sense of when or how they might use the method.

This problem extends to the ways books introduce examples of ideas, as they always give the most simple version. Exhibit 4.1 shows the answers students give to math problems in research studies, highlighting the nature of the problem caused by textbook questions.

EXHIBIT 4.1

Eleven-year-olds were shown the following figure and asked: Is line a parallel to line c?

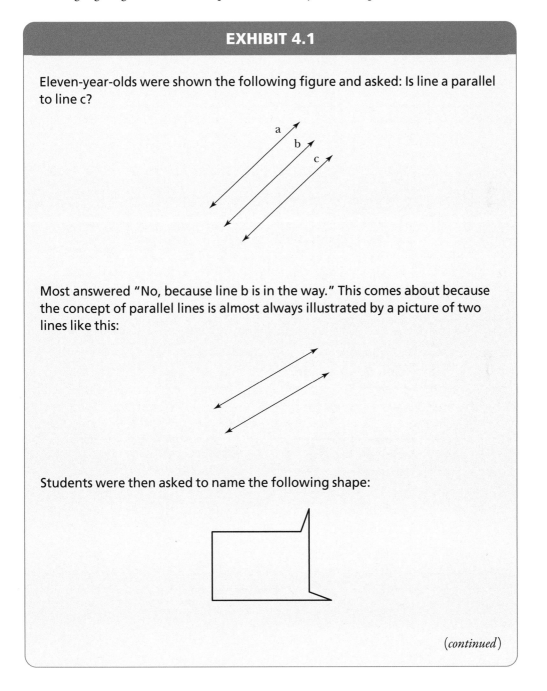

Most answered "No, because line b is in the way." This comes about because the concept of parallel lines is almost always illustrated by a picture of two lines like this:

Students were then asked to name the following shape:

(continued)

(*continued*)

Most were unable to. The shape *is* a hexagon (a six-sided polygon), but hexagons are almost always shown in this way:

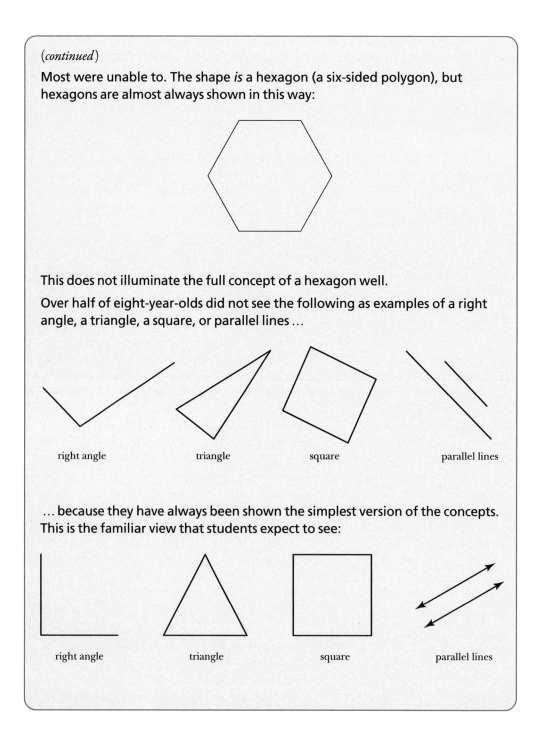

This does not illuminate the full concept of a hexagon well.

Over half of eight-year-olds did not see the following as examples of a right angle, a triangle, a square, or parallel lines …

right angle   triangle   square   parallel lines

… because they have always been shown the simplest version of the concepts. This is the familiar view that students expect to see:

right angle   triangle   square   parallel lines

The fact that over half of the students in the studies could not name the shapes tells us something important: when textbooks introduce only the simplest version of an idea, students are denied the opportunity to learn what the idea really is. Students were unable to name the different examples because the textbook authors had given "perfect examples" each time. When learning a definition, it is helpful to offer different examples—some of which barely meet the definition and some of which do not meet it at all—instead of perfect examples each time.

Mathematics teachers should also think about the width and breadth of the definition they are showing, and sometimes this is best highlighted by *non-examples*. When learning a definition, it is often very helpful to see both examples that fit the definition and others that *do not* fit, rather than just presenting a series of perfect examples. For example, when learning about birds it can be helpful to think about bats and why they are not birds, rather than to see more and more examples of sparrows and crows.

The misconceptions that students form when shown perfect examples is analogous to the problems students develop when practicing isolated methods over and over. Students are given uncomplicated situations that require the simple use of a procedure (or often, no situation at all). They learn the method, but when they are given realistic mathematics problems or when they need to use math in the world, they are unable to use the methods (Organisation for Economic Co-operation and Development, 2013). Real problems often require the choice and adaptation of methods that students have often never learned to use or even think about. In the next chapter we will look at the nature of rich and effective mathematics problems that avoid these problems.

In an award-winning research study in England, I followed students for three years through a practice approach to mathematics—students were shown isolated examples in math class, which they practiced over and over again (Boaler, 2002a). I contrasted this with an approach in which students were shown the complexity of mathematics and expected to think conceptually at all times, choosing, using, and applying methods. The two approaches were taught in separate schools, to students of the same background and achievement levels, both in low-income areas of the country. The students who were taught to practice methods over and over in a disciplined school in which there were high levels of "time on task" scored at significantly lower levels on the national mathematics examination than students who practiced much less but were encouraged to think conceptually. One significant problem the students from the traditional school faced in the national examination—a set of procedural questions—was that they did not know which method to choose to answer questions. They had practiced methods over and over but had never been asked to consider a situation and choose a method. Here are two of the students from the school reflecting on the difficulties they faced in the exam:

> It's stupid really 'cause when you're in the lesson, when you're doing work—even when it's hard—you get the odd one or two wrong, but most of them you get right and you think, "Well, when I go into the exam, I'm gonna get most of them right," 'cause you get all your chapters right. But you don't. (Alan, Amber Hill)

> It's different, and like the way it's there like—not the same. It doesn't like tell you it—the story, the question; it's not the same as in the books, the way the teacher works it out. (Gary, Amber Hill)

The oversimplification of mathematics and the practice of methods through isolated simplified procedures is part of the reason we have widespread failure in the United States and the United Kingdom. It is also part of the reason that students do not develop mathematical mindsets; they do not see their role as thinking and sense making; rather, they see it as taking methods and repeating them. Students are led to think there is no place for thinking in math class.

In a second study, conducted in the United States, we asked students in a similar practice model of math teaching what their role was in the math classroom (Boaler & Staples, 2005). A stunning 97% of students said the same thing: their role was to "pay careful attention." This passive act of watching—not thinking, reasoning, or sense making—does not lead to understanding or the development of a mathematical mindset.

Students are often given math practice as homework. There is a lot of evidence that homework, of any form, is unnecessary or damaging; I will share some of this evidence in Chapter Six. As a parent, I know that homework is the most common source of tears in our house, and the subject that is most stressful at home is math, especially when math homework consists of nothing but long lists of isolated questions.

Pages of practice problems are sent home—with no thought, it seems, of their negative effect on the home environment that evening. But there is hope: schools that decide to end homework see no reductions in students' achievement and significant increases in the quality of home life (Kohn, 2008).

Large research studies have shown that the presence or absence of homework has minimal or no effects on achievement (Challenge Success, 2012) and that homework leads to significant inequities (Program for International Student Assessment, 2015) (an issue I will return to in Chapter Six), yet homework plays a large negative role in the lives of many parents and children. Research also shows that the only time homework is effective is when students are given a worthwhile learning experience, not worksheets of practice problems, and when homework is seen not as a norm but as an occasional opportunity to offer a meaningful task. My daughters are in schools that know the research on homework and usually assign only worthwhile math homework, such as KENKEN puzzles, but occasionally teachers have sent home 40 questions of practice, on a method such as subtraction or multiplication. I've seen my children's spirits fall when they pull out this kind of homework. In those moments, I explain to them that the page of repeated questions is not what math really is, and after they have shown that they can solve a few of the questions—usually four or five—I suggest they stop. I write a note to the teacher saying that I am satisfied that they have understood the method, and I don't want them working on 35 more questions, as this would give them damaging ideas about the nature of math.

If you are working in a school where homework is required, there are homework problems to give that are much more productive than pages of math practice. Two innovative teachers I work with in Vista Unified School District, Yekaterina Milvidskaia and Tiana Tebelman, developed a set of homework reflection questions that they choose from each day to help their students process and understand the mathematics they have met that day at a deeper level. They typically assign one reflection question for students to respond to each night and one to five mathematical questions to work on (depending on the complexity of the problems). Exhibit 4.2 shows the reflection questions they have developed, from which they choose *one* each night.

Yekaterina and Tiana have used these reflection questions for two years and have noticed a really positive impact on their students, who now reflect on what they have learned in class, synthesize their ideas, and ask more questions in class.

Each year they have given their students a midyear survey to gather data and get their feedback on their classroom practices, including their new homework approach. When they asked students "Please provide us with feedback on your homework format this year," they received the following responses:

> I think that the way we do our homework is very helpful. When you spend more time reflecting about what we learned (written response), and less time doing more math (textbook), you learn a lot more.

> I feel like the homework questions help me reflect on what I learned from the day. If I do not quite remember something, then it gives me a chance to look back into my composition book.

> This year I really like how we do our homework. I understand how to do my homework because of the reflections; those really help me because then I can remember what I did in class that day.

> Having the reflection questions does actually help me a lot. I can see what I need to work on and what I'm doing good on.

The students talk about the ways the questions have helped them in their learning of mathematics. The questions are much less stressful for students, which is always important, and they invite students to think conceptually about big ideas, which is invaluable. Questions that ask students to think about errors or confusions are particularly helpful in encouraging students' self-reflection, and they will often result in the students' understanding the mathematics for the first time. Such questions also give the teacher really important information that can guide their teaching. Similar questions can be written for students at the end of class, as "exit tickets" that they write before they leave the lesson. I will share more ideas for reflection questions in Chapter Eight.

As I mentioned in Chapter One, the PISA team at the Organisation for Economic Co-operation and Development (OECD) not only give mathematics tests to students but also collect data on students' mindset beliefs and mathematics strategies. From looking at the strategies the 13 million students use, the data show that the lowest-scoring students in the world are those who use a memorization strategy. These students prepare for mathematics tests by trying to memorize methods. The highest-scoring students in the world are those who approach mathematics looking at and thinking about the big ideas and the connections between them. Figure 4.4 shows the achievement differences for students who use these different strategies.

One of the very best things we can do for students is to help them develop mathematical mindsets, whereby they believe that mathematics is about thinking, sense making, big ideas, and connections—not about the memorization of methods.

One excellent method for preparing students to think and learn in these ways—appreciating the connected, conceptual nature of mathematics—is a teaching strategy called "number talks."

Math Homework Reflection Questions
Part 1: Written Response Questions

*Your response to the question(s) chosen should be very detailed! Please write in complete sentences and be ready to share your response in class the next day.

1. What were the main mathematical concepts or ideas that you learned today or that we discussed in class today?

2. What questions do you still have about _____?
If you don't have a question, write a similar problem and solve it instead.

3. Describe a mistake or misconception that you or a classmate had in class today. What did you learn from this mistake or misconception?

4. How did you or your group approach today's problem or problem set? Was your approach successful? What did you learn from your approach?

Exhibit 4.2

This is also the very best strategy I know for teaching number sense and math facts at the same time. The method was developed by Ruth Parker and Kathy Richardson. This is an ideal short teaching activity that teachers can start lessons with or parents can use at home. It involves posing an abstract math problem and asking students to show how they solve the problem mentally. The teacher then collects the different methods students give and looks at why they work. For example, a teacher may pose 15×12 and find that students solve the problem in these five different ways:

$15 \times 10 = 150$	$30 \times 12 = 360$	$12 \times 15 =$	$12 \times 5 = 60$	$12 \times 12 = 144$
$15 \times 2 = 30$	$360 \div 2 = 180$	6×30	$12 \times 10 = 120$	$12 \times 3 = 36$
$150 + 30 = 180$		$6 \times 30 = 180$	$120 + 60 = 180$	$144 + 36 = 180$

5. Describe in detail how someone else in class approached a problem. How is their approach similar or different to the way you approached the problem?

6. What new vocabulary words or terms were introduced today? What do you believe each new word means? Give an example/picture of each word.

7. What was the big mathematical debate about in class today? What did you learn from the debate?

8. How is _____ similar or different to _____?

9. What would happen if you changed _____?

10. What were some of your strengths and weaknesses in this unit? What is your plan to improve in your areas of weakness?

Licensed under Creative Commons Attribution 3.0 by Yekaterina Milvidskaia and Tiana Tebelman

E x h i b i t 4 . 2

Students love to give their different strategies and are usually completely engaged and fascinated by the different methods that emerge. Students learn mental math, they have opportunities to memorize math facts, and they also develop conceptual understanding of numbers and of the arithmetic properties that are critical to success in algebra and beyond. Two books, one by Cathy Humphreys and Ruth Parker (Humphreys & Parker, 2015) and another by Sherry Parrish (Parrish, 2014), illustrate many different number talks to work on with secondary and elementary students, respectively. Number talks are also taught through a video on youcubed, which is an extract from my online teacher/parent course (http://www.youcubed.org/category/teaching-ideas/number-sense/).

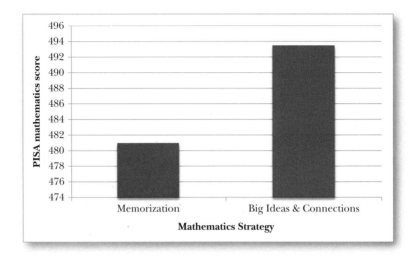

FIGURE 4.4   Mathematics strategies and achievement
Source: PISA, 2012.

Number talks are the best pedagogical method I know for developing number sense and helping students see the flexible and conceptual nature of math.

What about Older Students?

I have spent time in this chapter discussing the important pathways young children need to be set upon, through engaging with numbers conceptually, developing the idea that math is a subject that should make sense and that can be approached actively. It is ideal for students to take this pathway from the outset, but we know that anyone can change their pathway and their relationship with mathematics at any time. The next chapter will talk about middle school students and adults who hated and feared math, viewing it as a procedural subject. When they engaged in math differently—exploring the connections and patterns at the heart of the subject—and when they were given growth mindset messages about their potential, they changed completely. From that point on they approached mathematics differently and their learning pathways changed. I have seen this change happen in students of all ages, including the undergraduates I teach at Stanford. Figure 4.5 shows the impact of a mindset intervention given in the spring semester of seventh grade (Blackwell, Trzesniewski, & Dweck, 2007). Research tells us that students' relative achievement declines when they first move to middle school, but for students given a mindset intervention their decline was reversed.

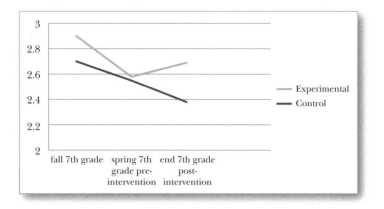

FIGURE 4.5 A mindset intervention
Source: Blackwell et al., 2007.

Mindset messages are very important for students; when these are accompanied by different mathematics opportunities, amazing things happen, at any age.

Math Apps and Games

Another way to give students opportunities to develop a thinking, conceptual approach to mathematics is to engage them in math apps and games that approach mathematics conceptually. The vast majority of all math apps and games are unhelpful, encouraging drill and rote memorization. In this section I highlight four apps and games that I regard as valuable, as they visually engage students in conceptual mathematics. I am an advisor to three of the four companies (Wuzzit Trouble, Mathbreakers, and Motion Math).

Wuzzit Trouble

Wuzzit Trouble, a game developed by Stanford mathematician Keith Devlin and his team, helps students gain an understanding of important mathematics—adding and subtracting, factors and multiples—at the same time as they develop number flexibility and problem-solving strategies. The aim of the game is to free a little creature from a trap by turning small cogs to rotate the wheel and release the keys (see Figure 4.6). The game progresses to really challenging puzzles (as seen in Figure 4.7), and different versions are available for specific mathematics topics and difficulty levels.

FIGURE 4.6   Wuzzit Trouble figures

FIGURE 4.7   Wuzzit Trouble puzzle

Wuzzit Trouble is produced by BrainQuake; it currently runs on iOS and Android and is available free to download at http://wuzzittrouble.com/.

Mathbreakers

Mathbreakers is a video game, targeted at elementary grade levels, in which avatars move around a world armed with numbers (see Figure 4.8). The game not only allows students to play with math, which is itself valuable, but also lets them act on numbers; for example, by cutting them in half if they need a smaller number to move over a bridge. I think of this game as similar to Minecraft but with numbers. It teaches number sense conceptually, in an engaging open environment.

Mathbreakers is produced by Imaginary Number Co.; it runs on Mac, Windows, and Linux, and as of this writing the list price is $25. Discounted educator pricing is also available (https://www.mathbreakers.com/).

Number Rack

Number Rack is an app, targeted at elementary grade levels, that models a learning tool called a Rekenrek, developed at the Freudenthal Institute in Holland (http://www.k-5mathteaching resources.com/Rekenrek.html). A Rekenrek shows 10 beads on a bar—multiple bars can be added for numbers greater than 10. Students and teachers can slide beads to show number relationships. There is a shield that can be used to cover sets of beads so students can find the missing number in a relationship. This is an easy-to-use tool for counting and working with 10 frames. Students can physically move the beads and work on number bonds and relationships conceptually.

FIGURE 4.8   The Mathbreakers game

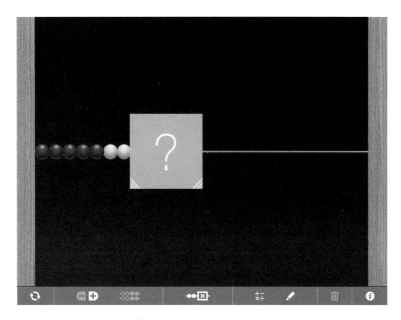

FIGURE 4.9   The Number Rack app

Number Rack (Figure 4.9), produced by the Math Learning Center, is available free on the Web (http://www.mathlearningcenter.org/web-apps/number-rack/).

Motion Math

Motion Math, targeted at elementary grade levels, offers a range of games that help students develop a visual understanding of important mathematics concepts—particularly numbers and fractions (see Figure 4.10). For example, in *Hungry Fish,* students merge numbers in order to feed a fish. *Hungry Fish* challenges players to find multiple pathways to create numbers. In *Pizza*, players operate a pizzeria, designing and selling pizzas; students learn proportional thinking, mental math, and even some economics. In the game *Fractions*, students move a bouncing ball containing a fraction to its correct place on the number line. In the *Cupcake* game students run their own business, making decisions about cupcakes, delivering them in their own vehicles, and making mathematical adjustments to their orders.

Motion Math, produced by Motion Math, runs on iOS and Android; individual games start at $2.99 (http://motionmathgames.com/).

There are other apps and games that help students develop number sense; I have chosen a few here to illustrate the important features to look for in good apps and games. All these different apps and games give insights into the mathematical concepts being learned, and they help students really *see* mathematical ideas.

FIGURE 4.10 Two Motion Math games: Hungry Fish and Pizza

Conclusion

New research on the brain tells us that the difference between successful and unsuccessful students is less about the content they learn and more about their mindsets. A growth mindset is important, but for this to inspire students to high levels of mathematics learning, they also need a mathematics mindset. We need students to have growth beliefs about themselves and accompany these with growth beliefs about the nature of mathematics and their role within it. With conceptual, investigative math teaching and mindset encouragement, students will learn to shed harmful ideas that math is about speed and memory, and that they either get it or they don't. This shift is key to mathematics achievement and enjoyment, and it can happen at any age, even for adults. This chapter has focused on what this means in the early years, particularly during the learning of numbers, but the ideas extend to all levels of math. Even math facts, one of the driest parts of math, can be taught conceptually and with sense making and understanding. As students are given interesting situations and encouraged to make sense of them, they will see mathematics differently, as not a closed, fixed body of knowledge but an open landscape that they can explore, asking questions and thinking about relationships. The next chapter will show some of the best ways to create this environment through rich and engaging mathematical tasks.

CHAPTER 5

Rich Mathematical Tasks

Teachers are the most important resource for students. They are the ones who can create exciting mathematics environments, give students the positive messages they need, and take any math task and make it one that piques students' curiosity and interest. Studies have shown that the teacher has a greater impact on student learning than any other variable (Darling-Hammond, 2000). But there is another critical part of the mathematics learning experience—in many ways, it is a teacher's best friend—and that is the curriculum teachers get to work with, the tasks and questions through which students learn mathematics. All teachers know that great mathematics tasks are a wonderful resource. They can make the difference between happy, inspired students and disengaged, unmotivated students. The tasks and questions used help develop mathematical mindsets and create the conditions for deep, connected understanding. This chapter will delve into the nature of true mathematics engagement and consider how it is brought about through the design of mathematics tasks.

   I have taught mathematics at middle school, high school, and undergraduate levels in England and the United States. I have also observed and researched hundreds of mathematics classrooms, across the K–16 level, in both countries, and studied students' learning of mathematics and the conditions that bring it about. I am fortunate to have had such a broad experience for many reasons, one of them being it has given me a great deal of insight into the nature of true mathematics engagement and deep learning. I have witnessed mathematical excitement, as it happens, with a range of different students, leading to the development of precious insights into mathematical ideas and relationships. Interestingly, I found that mathematics excitement looks exactly the same for struggling 11-year-olds as it does for high-flying students in top universities—it combines *curiosity, connection making, challenge,* and *creativity,* and usually involves *collaboration.* These, for me, are the 5 C's of mathematics engagement. In this chapter I will share what I have learned about the nature of mathematics engagement and excitement before considering the qualities of tasks that produce such engagement—tasks that all teachers can create in their own mathematics classrooms.

Rather than dissecting the nature of mathematics engagement in a clinical and abstract way, I want to introduce you to five cases of true mathematics excitement. I think of mathematics excitement as the pinnacle of mathematics engagement. These are all cases that I have personally witnessed among groups of people and that have given me important insights into the nature of the teaching and tasks that bring about such learning opportunities. The first case comes not from a school but from the unusual setting of a startup company in Silicon Valley. This case shows something powerful about mathematical excitement that I would love to capture and bottle for all teachers of mathematics.

Case 1. Seeing the Openness of Numbers

It was late December 2012, days before I was to fly to London for the holidays, when I first met Sebastian Thrun and his team at Udacity, a company producing online courses. I had been asked to visit Udacity to give the team advice about mathematics courses and ways to design effective learning opportunities. I walked into the airy space in Palo Alto that day and knew immediately that I had walked into a Silicon Valley startup—bikes were suspended on walls; young people, mostly men, wearing T-shirts and jeans, pored over computers or sat chatting about ideas; there were no office walls, only partitioned cubicles and light. I walked through the cubicles to the conference room at the back behind a glass wall. About 15 people had squeezed into the small room, sitting on chairs and the floor. Sebastian stepped forward and shook my hand, made some introductions, and invited me to sit down. He started firing questions at me: "What makes a good math course? How should we teach math? Why are students failing math?" He said that his friend Bill Gates had told him algebra was the reason we have widespread math failure in the United States. I cheekily replied, "Oh, Bill Gates the math educator told you that, did he?" His team members smiled, and Sebastian looked momentarily taken aback. He then asked, "Well, what do *you* think?" I told him that students were failing algebra not because algebra is so difficult, but because students don't have number sense, which is the foundation for algebra. Chris, one of the course designers who was also a former math teacher, nodded in agreement.

Sebastian continued firing questions at me. When he asked me what makes a good math question, I stopped the conversation and asked the group if I could ask *them* all a math question. They readily agreed, and I enacted a mini version of a number talk. I asked everyone to think about the answer to 18×5 and to show me, with a silent thumbs-up, when they had an answer. The thumbs started to pop up, and the team shared methods. There were at least six different methods shared that day, which I drew, visually, on the write-on table we sat around (see Figure 5.1).

We then discussed the ways the different methods were similar and different. As I drew the visual methods, the team members' eyes grew wider and wider. Some of them started to hop in their seats with excitement. Some shared that they had never imagined that there were so many ways to think about an abstract number problem; others said they were amazed that there was a visual image and it showed so much of the mathematics, so clearly.

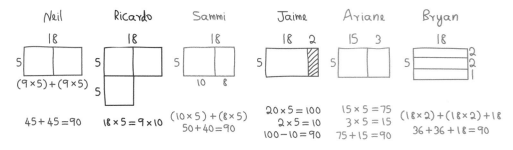

FIGURE 5.1 Visual solutions to 18 × 5

When I arrived in London, a few days later, I opened an email from Andy, the innovative young course designer at Udacity. He had made a mini online course on 18 × 5, which included going out on the street and interviewing passersby, collecting different methods. The team had been so excited by the ideas that they wanted to immediately put them out to the public, and they talked about making 18 × 5 T-shirts for everyone at Udacity to wear.

In the months following the meeting at Udacity, I met Luc Barthelet, then a director of Wolfram Alpha, one of the most important mathematics companies in the world. Luc had read about the different solutions to 18 × 5 I had shown in my book (Boaler, 2015a), which so excited him that he started asking everyone he met to solve 18 × 5. These reactions, these moments of intense mathematics excitement around an abstract number problem, seem important to share. How is it possible that these high-level math users, as well as young children, are so engaged by seeing and thinking about the different methods people use to solve a seemingly unexciting problem like 18 × 5? I propose that this engagement comes from people seeing the creativity in math and the different ways people *see* mathematical ideas. This is intrinsically interesting, but it's also true that most people I meet, even high-level mathematics users, have never realized numbers can be so open and number problems can be solved in so many ways. When this realization is combined with visual insights into the mathematical ways of working, engagement is intensified.

I have used this and similar problems with middle school students, Stanford undergrads, and CEOs of companies, all with equal engagement. I have learned through this that people are fascinated by flexibility and openness in mathematics. Mathematics is a subject that allows for precise thinking, but when that precise thinking is combined with creativity, flexibility, and multiplicity of ideas, the mathematics comes alive for people. Teachers can create such mathematical excitement in classrooms, with any task, by asking students for the different ways they see and can solve tasks and by encouraging discussion of different ways of seeing problems. In classrooms, teachers have to pay attention to classroom norms and teach students to listen to and respect each other's thinking; Chapter Seven will show a teaching strategy for this. When students have learned norms of respect and listening, it is incredible to see their engagement when different ways to solve a problem are shared.

Case 2. Growing Shapes: The Power of Visualization

The next case I want to share comes from a very different setting—a middle school classroom in a San Francisco Bay Area summer school where students had been referred because they were not performing well in the school year. I was teaching one of the four math classes with my graduate students at Stanford. We had decided to focus the classes on algebra, but not algebra as an end point, with students mindlessly solving for x. Instead, we taught algebra as a problem-solving tool that could be used to solve rich, engaging problems. The students had just finished sixth and seventh grades, and most of them hated math. Approximately half the students had received a D or an F in their previous school year (for more detail, see Boaler, 2015a; Boaler & Sengupta-Irving, 2015).

In developing a curriculum for the summer school, we drew upon a range of resources, including Mark Driscoll's books, Ruth Parker's mathematics problems, and two curricula from England—SMILE (which stands for secondary mathematics individualized learning experience) and Points of Departure. The task that created this case of mathematics excitement came from Ruth Parker; it asked the students to extend the growing pattern shown in Exhibit 5.1, made out of multilink cubes, to find how many cubes there would be in the 100th case. (Full-page task worksheets of all exhibits can be found in the Appendix.)

The students had multilink cubes to work with. In our teaching we invited the students to work together in groups to discuss ideas, sometimes groups we teachers chose, other times groups the students chose. On the day in question, I noticed an interesting grouping of three boys—three of the naughtiest boys in my class! They did not know each other before coming to the summer school, but all three spent most of the first week of summer school either off task or working to pull others off task. The boys would shout things out when others were at the board showing math and generally seemed more interested in social conversations than math conversations in the early days. Jorge had received an F in his last math class, Carlos a D, and Luke an A. But the day we gave the students this task, something changed. Incredibly, the three boys worked on this math task for 70 minutes, without ever stopping, becoming distracted, or moving off task. At one point some girls came over and poked them with pencils, which caused the boys to pick up their work and move to another table, they were so intensely engaged in the task and working to solve the problem.

All of our lessons were videotaped, and when we reviewed the film of the boys working that day we watched a rich conversation about number patterns, visual growth, and algebraic generalization. Part of the reason for the boys' intense engagement was an adaptation to the task that we had used—an adaptation that can be used with any math task. In classrooms, typically when function tasks such as the one we gave to the students are assigned, they are usually given with the instruction to find the 100th case (or some other high number) and ultimately the nth case. We did not start with this. Instead, we asked the students to first think alone, before moving to group work, about the ways they *saw* the shape growing. We encouraged them to think visually, not with numbers, and to sketch in their journals, showing us where they saw the extra cubes in each case. The boys saw the growth of the shape in different ways. Luke and Jorge saw the growth as cubes added to the bottom of the shape each time; this later became known by the class as

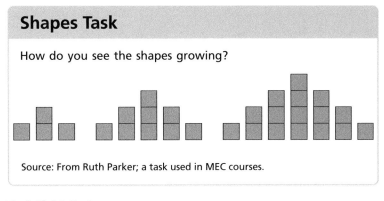

Exhibit 5.1

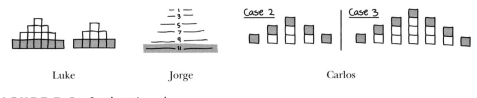

FIGURE 5.2 Students' work
Source: Selling, 2015.

the "bowling alley method," as the cubes arrived like a new line of pins in a bowling alley. Carlos saw the growth as cubes added to the top of the columns—what became known as the "raindrop method"—cubes dropping down from the sky, like raindrops, onto the columns (see Figure 5.2).

After the boys had spent time working out the function growth individually, they shared with each other their ideas for how the shape was growing, talking about where they saw the additional cubes in each case. Impressively, they connected their visual methods with the numbers in the shapes, not only working with their own methods but taking the time to explain the different methods to each other and using each other's methods. The three boys were intrigued by the function growth and worked hard to think about the 100th case, armed with their knowledge of the visual growth of the shape. They proposed ideas to each other, leaning across the table and pointing to their journal sketches. As is typical for mathematical problem solving, they zigzagged around, moving close to a solution, then further away, then back toward it again (Lakatos, 1976). They tried different pathways to the solution, and they broadly explored the mathematical terrain.

I have shown a video of the boys working to many conference audiences of teachers, and all have been highly impressed with the boys' motivation, perseverance, and high-level mathematical conversation. Teachers know that the perseverance shown by the three boys and the respectful ways they discussed each others' ideas, particularly in the context of summer school, is highly unusual, and they are curious as to how we were able to bring it about. They know that many students, particularly those who have been unsuccessful, give up when a task is hard and they don't immediately know the answer. That didn't happen in this case; when the boys were stuck, they looked back at their diagrams and tried out ideas with each other, many of which were incorrect but helpful in ultimately forming a pathway to the solution. After watching the case with teachers at conferences, I ask them what they see in the boys' interactions that could help us understand their high level of perseverance and engagement. Here are some important observations that reveal opportunities to improve the engagement of all students:

1) **The task is challenging but accessible**. All three boys could access the task, but it provided a challenge for them. It was at the perfect level for their thinking. It is very hard to find tasks that are perfect for all students, but when we open tasks and make them broader—when we make them what I refer to as "low floor, high ceiling"—this becomes possible for all students. The floor is low because anyone can see how the shape is growing, but the ceiling is high—the function the boys were exploring is a quadratic function whereby case n can be represented by $(n+1)^2$ blocks. We made the floor of the task lower by inviting the students to think visually about the case—although, as I will discuss later, this was not the only reason for this important adaptation.

2) **The boys saw the task as a puzzle**, they were curious about the solution, and they wanted to solve it. The question was not "real world" or about their lives, but it completely engaged them. This is the power of abstract mathematics when it involves open thinking and connection making.

3) **The visual thinking about the growth of the task gave the boys understanding of the way the pattern grew**. The boys could see that the task grew as a square of $(n+1)$ side length because of their visual exploration of the pattern growth. They were working to find a complex solution, but they were confident in doing so, as they had been given visual understanding to help them.

4) The boys were encouraged by the fact that **they had all developed their own way of seeing the pattern growth** and their different methods were valid and added different insights into the solution. The boys were excited to share their thinking with each other and use their own and each other's ideas in the solving of the problem.

5) **The classroom had been set up to encourage students to propose ideas without being afraid of making mistakes**. This enabled the boys to keep going when they were "stuck," by providing ideas, right or wrong, that enabled the conversation to continue.

6) We had taught the students to **respect each other's thinking.** We did this by valuing the breadth of thinking everyone could offer, not just the procedural thinking that some could offer, valuing the different ways people saw problems and made connections.

7) The students were **using their own ideas,** not just following a method from a book as they learned core algebraic content. The fact that they had proposed different visual ideas for the growth of the function made them more invested and interested in the task.

8) **The boys were working together**; the video shows clearly the way the boys built understanding through the different ideas they shared in conversation, which also enhanced their enjoyment of the mathematics.

9) The boys were working heterogeneously. Viewers of the video note that each boy offers something different and important. The high achiever keeps shouting out number guesses—something that may have been a successful strategy with more procedural questions—but the lower-achieving boys push him to think visually and ultimately more conceptually, and it is the combination of the different boys' thinking that ultimately helps them and leads to success.

Typically, growth pattern tasks are given to students with a numerical question such as "How many cubes are in the 100th case?" and "How many cubes are in the nth case?" We also asked students these questions, but we prefaced them with individual time in which students considered the visual growth of the shape. That changed everything. People think about the growth of the shape in many different ways, as shown in Figures 5.3 through 5.10. When we don't ask students to think visually, we miss an incredible opportunity to increase their understanding. These are some of the ways teachers and students I have worked with see the growth of the shape, accompanied by names they use to capture the growth.

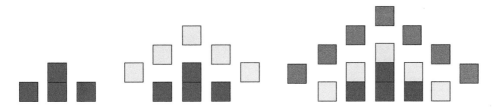

FIGURE 5.3 The Raindrop Method—cubes come from the sky like raindrops

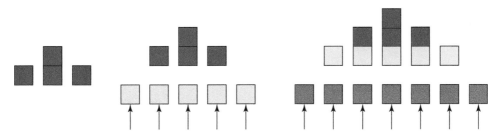

FIGURE 5.4 The Bowling Alley Method—cubes are added like pins in a bowling alley

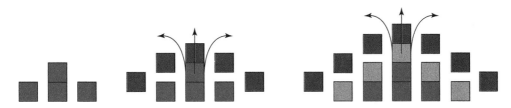

FIGURE 5.5 The Volcano Method—the middle column of cubes grows high and the rest follow like lava erupting from a volcano

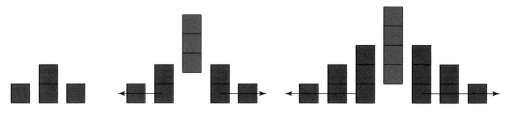

FIGURE 5.6   The Parting of the Red Sea Method—the columns part and the middle column arrives

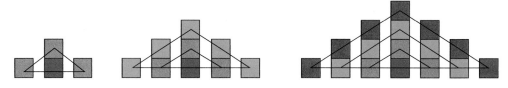

FIGURE 5.7   The Similar Triangles Method—the layers can be seen as triangles

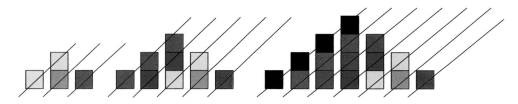

FIGURE 5.8   The Slicing Method—the layers can be viewed diagonally

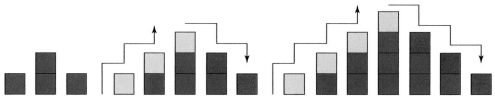

FIGURE 5.9   "Stairway to Heaven, Access Denied"—from *Wayne's World*

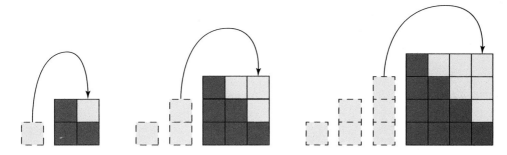

FIGURE 5.10 The Square Method—the shapes can be rearranged as a square each time

I recently gave this pattern-growth task to a group of high school teachers who did not take the time to explore the visual growth of the shape and instead produced a table of values like this:

case	#cubes
1	4
2	9
3	16
n	$(n+1)^2$

When I asked the teachers to tell me why the function was growing as a square, why it was $(n+1)$ *squared*, they had no idea. But this is why we see a squared function: the shape grows as a square, with a side of $(n+1)$, where n is the case number (see Figure 5.11).

When we do not ask students to think visually about the growth of the shape, they do not have access to important understanding about functional growth. They often cannot say what "n" means or represents, and algebra remains a mystery to them—a set of abstract letters they

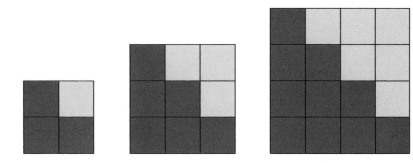

FIGURE 5.11 The Square Method 2

move around on a page. Our summer school students knew what "n" represented, because they had drawn it for themselves. They knew why the function grew as a square and why the nth case was represented by $(n+1)^2$. The algebraic expression they ultimately produced was meaningful to them. Additionally, students did not think they were finding a standard answer for us; they thought they were exploring methods and using their own ideas and thoughts, which included their own ways of seeing mathematical growth. In the final section of this chapter, I will review the ways these features of this task can be used in other tasks to produce increased student engagement and understanding.

Case 3. A Time to Tell?

When I share open, inquiry-based mathematics tasks with teachers, such as the growing shapes or "raindrop" task just discussed, they often ask questions such as, "I get that these tasks are engaging and create good mathematical discussions, but how do students learn new knowledge, such as trig functions? Or how to factorize? They can't discover it." This is a reasonable question, and we do have important research knowledge about this issue. It is true that while ideal mathematics discussions are those in which students use mathematical methods and ideas to solve problems, there are times when teachers needs to introduce students to new methods and ideas. In the vast majority of mathematics classrooms, this happens in a routine of teachers showing methods to students, which students then practice through textbook questions. In better mathematics classrooms, students go beyond practicing isolated methods and use them to solve applied problems, but the order remains—teachers show methods, then students use them.

In an important study, researchers compared three ways of teaching mathematics (Schwartz & Bransford, 1998). The first was the method used across the United States: the teacher showed methods, the students then solved problems with them. In the second, the students were left to discover methods through exploration. The third was a reversal of the typical sequence: the students were first given applied problems to work on, even before they knew how to solve them; then they were shown methods. It was this third group of students who performed at significantly higher levels compared to the other two groups. The researchers found that when students were given problems to solve, and they did not know methods to solve them, but they were given opportunity to explore the problems, they became curious, and their brains were primed to learn new methods, so that when teachers taught the methods, students paid greater attention to them and were more motivated to learn them. The researchers published their results with the title "A Time for Telling," and they argued that the question is not "Should we *tell* or explain methods?" but "When is the best time do this?" Their study showed clearly that the best time was *after* students had explored the problems.

How does this work in a classroom? How do teachers give students problems that they cannot solve without the students experiencing frustration? In describing how this works in practice, I will draw from two different cases of teaching.

The first example comes from the research study I conducted in England that showed that students who learned mathematics through a project-based approach achieved at significantly higher levels in mathematics, both in standardized tests (Boaler, 1998) and later in life (Boaler, 2005), than students who worked traditionally. In one of the tasks I observed in the project school, a group of 13-year-old students were told that a farmer wanted to make the largest enclosure she could out of 36 1-meter pieces of fencing. The students set about investigating ways to find the maximum area. Students tried different shapes, such as squares, rectangles, and triangles, and tried to find a shape with the biggest possible area. Two students realized that the biggest area would come from a 36-sided shape, and they set out to determine the exact area (see Figure 5.12).

They had divided their shape into 36 triangles, and they knew the base of each triangle was 1 meter and the angle at the vertex was 10 degrees (see Figure 5.13).

However, this alone was not enough to find the area of the triangle. At this point the teacher of the class showed the students trigonometry and the ways that a sine function could be used to give them the height of the triangle. The students were thrilled to learn this method, as it helped them solve the problem. I watched as one boy excitedly taught his group members how to use a sine function, telling them he had learned something "really cool" from the teacher. I then remembered the contrasting lesson I had watched in the traditional school a week earlier, in which the teacher had given the students trig functions and then pages of questions to practice them on. In that case, the students had thought the trig functions were extremely boring and unrelated to their lives. In the project school, the students were excited to learn about

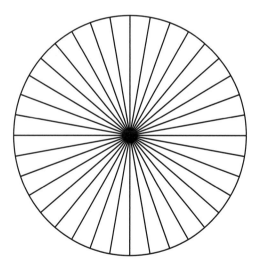

FIGURE 5.12 A 36-sided fence yields the largest enclosure area

FIGURE 5.13 The triangle formed from a 1-meter fence section

trigonometry and saw the methods as "cool" and useful. This heightened motivation means that the students learned the methods more deeply, and it is a large part of the success of the students in the project school in examinations and life.

The second example of students learning methods after they were given problems came from a research study I conducted in the United States. Similar to the UK study, it showed that students learned at significantly higher levels when taught mathematics through a conceptual approach focused on connections and communications (Boaler & Staples, 2005). More detail on both school approaches is given in my book *What's Math Got to Do with It?* (Boaler, 2015a). One day I was in one of the pre-calculus classrooms of the successful school, which I called Railside, when the teacher taught a lesson focused on finding the volume of a complex shape. The teacher, Laura Evans, was preparing the students to learn calculus and to find an area under a curve using integrals, but she did not, as typically happens, teach the formal method to the students first. Instead, she gave students a problem that needed this knowledge and asked them to think about how they would solve it. The problem was to work out a way to find the volume of a lemon. To think about this, she gave each group a lemon and a large knife and asked them to explore possible solutions (see Figure 5.14).

FIGURE 5.14 What is the volume of a lemon?
Source: Shutterstock, Copyright: ampFotoStudio.

After groups had discussed the problem, different students came to the board to excitedly share their ideas. One group had decided to plunge the lemon into a bowl of water to measure the displacement of the water. Another had decided to carefully measure the size of the lemon. A third had decided to cut the lemon into thin slices and think of the slices as two-dimensional sections, which they divided into strips, getting close to the formal method for finding the area under a curve that is taught in calculus (see Figure 5.15).

When the teacher explained to students the method of using integrals, they were excited and saw the method as a powerful tool.

In both of these cases the order of teaching methods was reversed. The students learned trig methods and limits *after* exploring a problem and encountering the need for the methods. The teacher taught them the methods when they were needed, rather than the usual approach of teaching a method that students then practiced. This made a world of difference to the students' interest in and subsequent understanding of the methods.

As I recalled in Chapter Four, Sebastian Thrun explained to me the key role played by intuition in his mathematical work. He said that he did not make mathematical progress unless he intuitively felt it was the right direction. Mathematicians also highlight the critical role played by intuition in their work. Leone Burton interviewed 70 research mathematicians and found that 58 of them embraced and talked about the essential role of intuition in their work (Burton, 1999). Hersh draws a similar conclusion when studying mathematical work: "If we look at mathematical

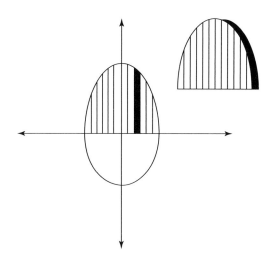

FIGURE 5.15   Calculating a lemon's volume by sections

practice, the intuitive is everywhere" (Hersh, 1999). So why is intuition, so central to mathematics, absent in most math classrooms? Most children do not think intuition is even allowed in their math work. When students were asked to think about finding the volume of a lemon, they were asked to think intuitively about math. Many mathematics problems could be posed to students with the request to think intuitively about ways to solve the problem. Young children could be given different triangles and rectangles and asked to think about ways they might find the area of a triangle, *before* being told an area formula. Students could think about ways to capture the differences between data sets *before* being taught about mean, mode, or range. Students can explore relationships in circles before being told the value of pi. In all cases, when students go on to learn the formal methods, their learning will be deeper and more meaningful. When students are asked to think intuitively, many good things happen. First, they stop thinking narrowly about single methods and consider mathematics more broadly. Second, they realize they have to use their own minds—thinking, sense making, and reasoning. They stop thinking their task is just to repeat methods, and they realize their task is to think about the appropriateness of different methods. And third, as the Schwartz and Bransford research study showed, their brains become primed to learn new methods (Schwartz & Bransford, 1998).

Case 4. Seeing a Mathematical Connection for the First Time (Pascal's Triangle)

My next case of mathematics excitement comes from a professional development workshop I was observing. The teacher of the workshop was Ruth Parker, an amazing educator who offers workshops for teachers that give them access to mathematical understandings they have never had before. I chose this case to share because I saw something that day that I have seen many times since: a task that allowed a teacher, Elizabeth, to see a mathematical connection so powerful it made her cry. The teacher was an elementary teacher who, like many others, had thought of mathematics as a set of procedures to follow. She did not know that mathematics was a subject full of rich connections. It is not uncommon for people who have always believed that mathematics is a disconnected set of procedures to be extremely moved when they see the rich connections that make up mathematics.

Ruth's workshop, similar to our summer school teaching, was focused on algebraic thinking, and she engaged the teachers in many function pattern tasks. The task chosen by Ruth that day was a lovely low floor, high ceiling task that appears simple but leads to wonderful complexity. The Cuisenaire Rod Train Task is shown in Exhibit 5.2. For the teachers in Ruth's workshop, it led to explorations of exponential growth and negative exponents.

Elizabeth and the other teachers set to work on the task, ordering and arranging Cuisenaire rods to find all the ways they could make trains the length of three rods they had chosen. Some teachers in the workshop chose to start with the 10 rod, which made their task extremely difficult—there are 512 ways to make trains that are as long as the 10 rod! Ruth knew that her role as the teacher was not to rescue them, but to let them wade around inside the mathematics of the problem. After a lot of struggle, some of them remembered something they had learned earlier in the workshop—a key mathematical practice students can go through 11 years of school without learning: to try a smaller case. The teachers worked on different rod lengths and started to see a pattern emerging, both visually and numerically (see Exhibit 5.2).

At this point Ruth introduced the teachers to Pascal's triangle, and asked teachers to search for the connection between the Cuisenaire rods problem and the famous triangle (see Exhibit 5.3 and Figure 5.16).

After much struggle the teachers then, with some amazement, saw that their Cuisenaire train combinations were all inside Pascal's triangle. This was the moment when Elizabeth was moved to tears—an emotion I fully understand. For anyone who has only ever seen mathematics as a set of disconnected procedures, who then gets the opportunity to explore visual and numerical patterns, seeing and understanding connections, the experience is powerful. Elizabeth gained an intellectual empowerment in those moments that told her that she could, herself, discover mathematical insights and connections. From that point on, Elizabeth's relationship with math changed, and she never looked back. I caught up with Elizabeth a year later, when she was retaking Ruth's course to experience more powerful mathematics learning, and she told me all of the wonderful ways she had changed her math teaching and the new mathematical excitement that she was seeing in her third-grade students.

Elizabeth's experience of seeing mathematics in an entirely new way when she was introduced to mathematical connections is one I have seen repeated over and over again with different children and adults. The strength of emotions I see relates directly to the experience of seeing, exploring, and understanding mathematical connections.

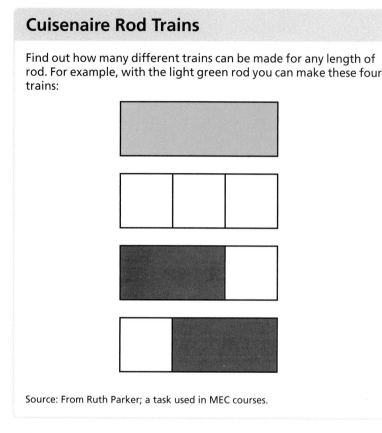

Cuisenaire Rod Trains

Find out how many different trains can be made for any length of rod. For example, with the light green rod you can make these four trains:

Source: From Ruth Parker; a task used in MEC courses.

Exhibit 5.2

Case 5. The Wonders of Negative Space

This case draws from a task I have used with my teacher education group at Stanford and with different groups of teachers; it creates such intensity of excitement that it seems important to share. The task is again a pattern-growth task but with an added twist; it is that added twist I want to focus on. The pattern-growth task came from Carlos Cabana, a wonderful teacher I work with. Exhibit 5.4 shows the two questions he usually gives to students.

One of the questions posed in the task is how many tiles there would be in Figure −1; in other words, if the pattern was to extend backward to case 1, case 0, and case −1, how many tiles would there be in case −1? In giving this task to teachers, I found that the number of tiles was an easy question for them to answer; what was much more interesting and challenging was the question

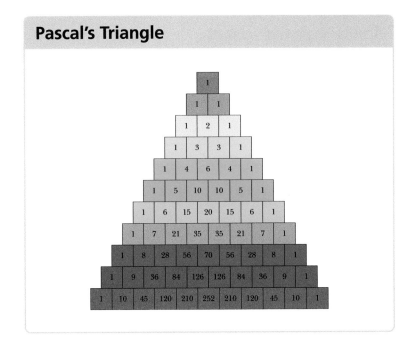

Exhibit 5.3

FIGURE 5.16 Pascal's triangle in Cuisenaire rods

Negative Space Task

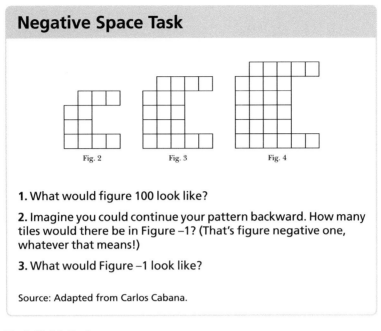

Fig. 2   Fig. 3   Fig. 4

1. What would figure 100 look like?

2. Imagine you could continue your pattern backward. How many tiles would there be in Figure –1? (That's figure negative one, whatever that means!)

3. What would Figure –1 look like?

Source: Adapted from Carlos Cabana.

E x h i b i t   5 . 4

of what the negative-one case would look like. When I added this question to the task, amazing things happened. First of all, the solution (which I won't give away) is challenging, and teachers laughed about their heads hurting and synapses firing when trying to figure it out. There is more than one way to get to the negative-one case, and not only is there more than one correct visualization, but there is also more than one numerical solution, because the question moves into unusually uncharted and exciting waters: considering the question of what negative squares look like. Some of the teachers discovered that they would need to think about *negative space,* and what a tile would look like if it was inverted onto itself. When I gave my teacher education group this task at Stanford, they were jumping over tables with excitement trying to represent negative space, poking holes through the paper to show the tiles going into negative space. One of the teachers realized and shared that the function could be represented as a parabola on a graph (see Figure 5.17). Another teacher asked where the parabola would go—would it stay on the positive y axis or flip below the axis?

This question was extremely engaging for the group, and they excitedly tried to work it out. At the end of the session, the teacher candidates reflected that they had now experienced true mathematics excitement and knew what they wanted students in their classes to experience.

But what caused this excitement, which I have now seen repeated in many different places? When I gave this task to a group of teacher leaders in Canada recently, they were so engaged by the task that I could not get them to stop working on it, which they laughed about. People tweeted "Jo Boaler can't bring us back from the task she has set."

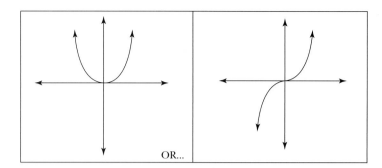

FIGURE 5.17   Parabola Dilemma

One reason the task is so exciting is that it involves thinking about negative space, stepping into another dimension—and that is exciting, period. This is something that mathematics allows us to do, and one reason why mathematics is an exciting subject. Additionally, the students believed that they were exploring uncharted waters; they were not finding an answer to a question that the textbook and teacher knew, and that increased their excitement dramatically. When students were asking about the direction of the parabola, they felt that they could ask anything—that mathematics was an open subject and that when they discovered a new idea (a parabola) they could take it further with another question that they posed. The visual representation of the mathematical pattern was again hugely important for the levels of engagement.

Before reflecting on what these different cases of intense mathematics excitement mean for the design of engaging tasks, I want to introduce one last case, this one from a third-grade classroom.

Case 6. From Math Facts to Math Excitement

Chapter Four discussed the importance of teachers' changing the ways they encourage students to learn math facts, moving from activities that are often traumatic for students—timed tests of isolated facts and hours of memorization—to engaging activities that support important brain connections. To help teachers make such changes, I wrote a paper with my Youcubed colleagues, as I described in the last chapter, entitled "Fluency without Fear," and I posted the paper on our site in the hope that it would reach many teachers, but we could not have anticipated the eventual extent of the impact, with major newspapers across the United States covering the ideas in the paper. One activity that we gave to teachers created a positive impact in a different way, as teachers circulated it widely among themselves using various social media, alongside photos of their students enjoying the activity and making important brain connections.

The activity (described in the preceding chapter) that proved to be so important and popular was a game called "How close to 100?"

   One of the teachers who took my online class and subsequently changed her mathematics teaching was RoseAnn Hernandez, a third-grade teacher in a Title 1 school—that is, a school in California where at least 40% of the children are from low-income families. RoseAnn has a copy of Youcubed's seven positive math norms (see Chapter Nine) displayed on her wall for all the students to see. RoseAnn shared with me her students' excitement in playing the game, as well as the important mathematical opportunities that they received (see Figure 5.18). RoseAnn is a very thoughtful teacher; she not only gave the game to students but also prepared them with a discussion before the game; she also prepared extension activities for any students who were working faster. Before the game she asked students to think about the ways dice can be used as a math tool. She gave them opportunities to roll two dice and take turns stating the multiplication factors and products that they generated. She then asked them an important question: How are multiplication and area related? The students thought carefully about this. RoseAnn then invited students to work on the game in pairs and to think about what they were learning as they played. She also challenged the students to decompose the numbers and discover different ways to write the number sentences on the back of their paper if they finished early. The students played the game with much excitement, and when RoseAnn asked them to rate their enjoyment on a 1-to-5 scale, 95% of the students gave it the highest possible rank of 5.

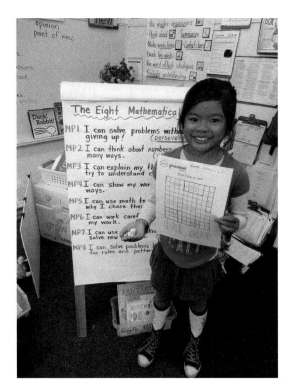

FIGURE 5.18 Third-grade student completes "How close to 100?"

The following are some of the students' important words as they reflected on the game:

"It challenged me to make my brain think."

"It was a fun way to explore math and learn."

"It gave me a lot of practice with multiplication."

"It's a fun way to learn multiplication facts."

"I learned that multiplication and area are related."

"I know now how division, multiplication, and area are all related because I can see it!"

The level of students' excitement in playing the game was matched only by the power of the mathematics they learned. It is noteworthy that as much as the students enjoyed the game, their comments all talked about the mathematics they learned. The students engaged in visual and numerical thinking about multiplication, division, and area, learning math facts through enjoyment and deep engagement—a far cry from the memorization of times tables.

In all of these six cases of mathematics excitement, the mathematics task was central (and supported by important teaching). The next section will review the important design aspects of these six tasks that can be applied to all mathematics tasks, regardless of grade level. It is also important to note that in all six cases students were working with each other, sometimes thinking alone but often collaborating on ideas, in classrooms where they were given positive growth mindset messages. I will now turn to the ways we can build these important design elements into any mathematics task.

From Cases of Mathematics Excitement to the Design of Tasks

We are emerging from an unproductive period in education. Since the No Child Left Behind Act was introduced by the Bush administration, teachers started to come under pressure to use "scripted" curriculum and pacing guides, even though they knew they were damaging their students. Many teachers felt deprofessionalized by this; they felt that important teaching decisions had been taken out of their hands. Fortunately, this time is ending; we are entering a much more positive time, with teachers being trusted to make important professional decisions. One of the aspects of teaching for a mathematics mindset that I am most excited about is the transformations that we can make in mathematics classrooms through giving important messages and opening up mathematics tasks. This opening up of tasks gives students the space to learn and is absolutely essential in developing mathematical mindsets.

A range of rich, open tasks are now available to teachers through websites, which I will list at the end of this chapter. But many teachers do not have time to search through websites.

Fortunately, teachers don't need to find new curriculum materials, as they can make adaptations to the tasks in the curriculum they use, opening them to create new and better opportunities for students. To do this, teachers may need to develop their own new mindsets as designers—that is, as people who can introduce a new idea and create new, enhanced learning experiences. The mathematical excitement I described earlier came, in a number of cases, from adapting a familiar task. In the growing shapes task, for example, the simple instruction for students to visualize the shapes growing changed everything, giving students access to understandings that would not have been possible otherwise. When teachers are designers, creating and adapting tasks, they are the most powerful teachers they can be. Any teacher can do this; it does not require special training. It involves knowing about the qualities of positive math tasks and approaching tasks with the mindset to improve them.

In designing and adapting math tasks for better learning, there are six questions that, if asked and acted upon in the task, increase their power incredibly. Some tasks are more suited to some questions than others, and many are naturally combined, but I am confident in saying every task will be made richer by paying attention to at least one of the following six questions.

1. Can You Open the Task to Encourage Multiple Methods, Pathways, and Representations?

There is nothing more important that teachers can do with tasks than to open them up so students are encouraged to think about different methods, pathways, and representations. When we open a task we transform its learning potential. Opening can happen in many ways. Adding a visual requirement, such as those shown in the growing shapes and negative space tasks, is a great strategy. Another way to open a task that is extremely mathematically productive is to ask students to make sense of their solutions.

Cathy Humphreys is wonderful teacher. In a book we coauthored, we show six video cases of Cathy teaching her seventh-grade class, accompanied by her lesson plans. One of the videos shows Cathy asking the students to solve: 1 divided by 2/3. This could be a closed, fixed mindset question with one right answer and one method, but Cathy transforms the task by adding two requirements: that students make sense of their solution and that they offer a visual proof (see Figure 5.19). She starts the lesson by saying "You may know a rule for solving this question, but the rule doesn't matter today, I want you to make sense of your answer, to explain why your solution *makes sense*."

In the video case we see that some students thought the answer was 6, because you can manipulate the set of numbers (1, 2, and 3), with no mathematical sense making, and make 6. But they struggled to show this visually or make sense of it. Others were able to show, in a range of different visual representations, why there were one and a half 2/3's "inside 1." The requirement for students to show their thinking visually and make sense of their answers transformed the question from a fixed to a growth mindset task, and created a wonderful lesson, filled with sense making and understanding.

FIGURE 5.19   Students share solutions to 1 ÷ 2/3

2. Can You Make It an Inquiry Task?

When students think their role is not to reproduce a method but to come up with an idea, everything changes (Duckworth, 1991). The same mathematics content can be taught with questions that ask for a procedure or as questions that ask for students to think about ideas and use a procedure. For example, instead of asking students to find the area of a 12 by 4 rectangle, ask them how many rectangles they can find with an area of 24. This small adaptation changes students' motivation and understanding. In the inquiry version of the task, students use the formula for the area of a rectangle, but they also need to think about spatial dimensions and relationships, and what happens when one dimension changes (see Figure 5.20). The mathematics is more complex and exciting because students are using their ideas and thoughts.

Instead of asking students to name quadrilaterals with different qualities, ask them to come up with their own, as shown in Exhibit 5.5.

Another excellent task is the four 4's (see Exhibit 5.6). In this task you ask students to make all the numbers between 1 and 20 using four 4's and any operation; for example:

$$\sqrt{4} + \sqrt{4} + 4/4 = 5$$

This is a great activity for practicing operations, but it does not look like a practicing operations task, because the operations are beautifully embedded inside an inquiry task. When we posted

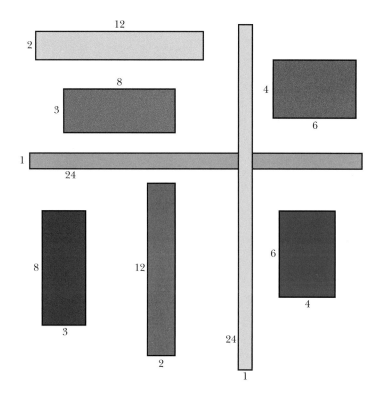

FIGURE 5.20   Rectangles with an area of 24

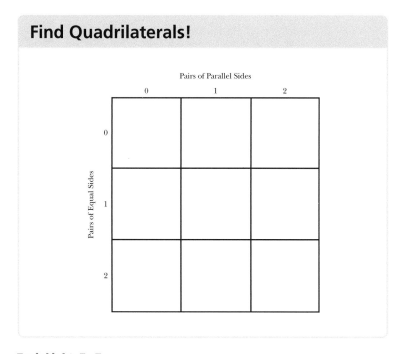

Exhibit 5.5

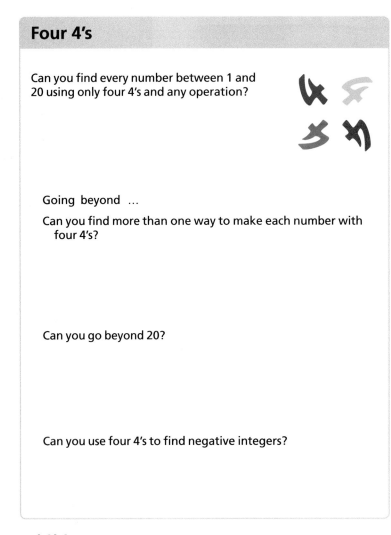

Four 4's

Can you find every number between 1 and 20 using only four 4's and any operation?

Going beyond …

Can you find more than one way to make each number with four 4's?

Can you go beyond 20?

Can you use four 4's to find negative integers?

Exhibit 5.6

this task on Youcubed.org, teachers told us the task was incredible. Here are two comments from Youcubed teachers:

"My students were so inspired &excited with the four 4's they decided to investigate three 3's, and the sky was the limit."

"The fours problem was amazing! I used it in my sixth-grade math class, and students were creating equations that led to discussions about distributive property, order of operations, variables … it was fantastic!"

(The full task on Youcubed includes advice on ways to introduce the task and organize students; see https://www.youcubed.org/wim-day-1/.)

Another way to open a task and make it an inquiry task is to ask students to write a magazine article, a newsletter, or a short book about it. This structure can work with any content. At Railside, in ninth grade the students were asked to write a book on y = mx + b; they filled pages showing what this meant, how it could look visually, situations in which it could be used, and their ideas on the meaning of the equation. In a high school geometry unit that three of my graduate students at Stanford (Dan Meyer, Sarah Kate Selling and Kathy Sun) created with me, we asked students to write a newsletter on similarity, using photos, tasks, cartoons, and any other media they wanted to show what they knew about the topic (see https://www.youcubed.org/wp-content/uploads/The-Sunblocker1.pdf). Exhibit 5.7 is a general form of the newsletter assignment we gave out.

3. Can You Ask the Problem Before Teaching the Method?

When we pose problems for which students need to know a method before we introduce the method, we offer a great opportunity for learning and for using intuition. The tasks described earlier that exemplified this were the finding the largest enclosure area for a fence task and the finding the volume of a lemon task. But this design component can be used with any area of mathematics—in particular, for any teaching of a standard method or formula, such as the area of shapes, the teaching of pi, and statistical formulas such as mean, mode, range, and standard deviation. Exhibit 5.8 shows an example.

After students have worked out their own ways of finding averages and discussed them as groups and as a class, they could be taught the formal methods of mean, mode, and range.

4. Can You Add a Visual Component?

Visual understanding is incredibly powerful for students, adding a whole new level of understanding, as we saw in the growing shapes task. This can be provided through diagrams but also through physical objects, such as multilink cubes and algebra tiles. I spent my early years growing up with Cuisenaire rods, as my mother was training to be an elementary teacher. I spent many happy hours playing with the rods, ordering them and investigating mathematical patterns. In an online course designed to give students important mathematics strategies, I teach students to draw any mathematics problem or idea (see https://class.stanford.edu/courses/Education/EDUC115-S/Spring2014/about). Drawing is a powerful tool for mathematicians and mathematical problem solvers, most of whom draw any problem they are given. When students are stuck in math class, I often ask them to draw the problem out.

Newsletter

You are writing a newsletter to share your learning on this mathematics topic with your family and friends. You'll have the chance to describe your understanding of the ideas and write about why the mathematical ideas you have learned are important. You'll also describe a couple of activities that you worked on that were interesting to you.

In creating your newsletter, you can draw on the following resources:

- Photos of different activities
- Sketches
- Cartoons
- Interviews/surveys

To refresh your memory, here are some of the activities we've worked on:

Please prepare the following four sections. You can change the titles of the sections to fit your work.

Headline News	New Discoveries
Explain the big idea of the mathematics and what it means in at least two different ways. Use words, diagrams, pictures, numbers, and equations.	Choose at least two different activities from the work we have done that helped you understand the concepts. For each activity: • Explain why you chose the activity. • Explain what you learned about through the activity. • Explain what was challenging about the activity. • Explain the strategies you used to address your challenge.
Connections	**The Future**
Choose one additional activity that helped you learn a mathematical idea or process that you can connect to some other learning. • Explain why you chose the activity. • Explain the big mathematical idea you learned from the activity. • Explain what you connected this idea to and how you see the connection. • Explain the importance of the connection and how you might use this in the future.	Write a summary for the newsletter that addresses the following: • What is the big mathematical idea useful for? • What questions do you still have about the big idea?

Exhibit 5.7

The Long Jump

You are going to try out for the long jump team, for which you need an average jump of 5.2 meters. The coach says she will look at your best jump each day of the week and average them out. These are the five jumps you recorded that week:

	Meters
Monday	5.2
Tuesday	5.2
Wednesday	5.3
Thursday	5.4
Friday	4.4

Unfortunately, Friday's was a low score because you weren't feeling that well!

How could you work out an average that you think would fairly represent your jumping? Work out some averages in different ways and see which you think is most fair, then give an argument for why you think it is fairest. Explain your method and try and convince someone that your approach is best.

Exhibit 5.8

Railside School, the highly successful school I studied, asked students to show connections through color coding. For example, when teaching algebra, they asked students to show functional relationships in many forms: as an expression, as a picture, in words, and on a graph. Many schools ask for these different representations. Railside was unusual in that they asked the students to show relationships in color; for example, to show the x in the same color in an expression, on the graph, and in the diagram. Chapter Seven, which describes the Railside approach in more detail, shows one of their color coding tasks. In other topic areas—for example, when asking students to identify congruent, vertical, and supplementary angles—you could also ask them to color and write about as many relationships as they can, using color to highlight the relationships. Exhibit 5.9 and Figure 5.21 show an example.

Further examples of color coding are given in Chapter Nine.

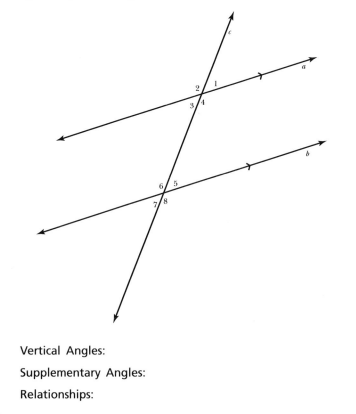

Parallel Lines and a Transversal

1. Use color coding to identify congruent angles.

2. Identify vertical and supplementary angles.

3. Write about the relationships you see. Use the color from your diagram in your writing.

Vertical Angles:

Supplementary Angles:

Relationships:

E x h i b i t   5 . 9

5. Can You Make It Low Floor and High Ceiling?

All of the preceding problems are low floor and high ceiling. The breadth of the space inside them means that they are accessible to a wide range of students and they extend to high levels.

One way to make the floor lower is to always ask students how they see a problem. This is an excellent question for other reasons too, as I have explained.

A great strategy for making a task higher ceiling is to ask students who have finished a question to write a new question that is similar but more difficult. When we were teaching a group of

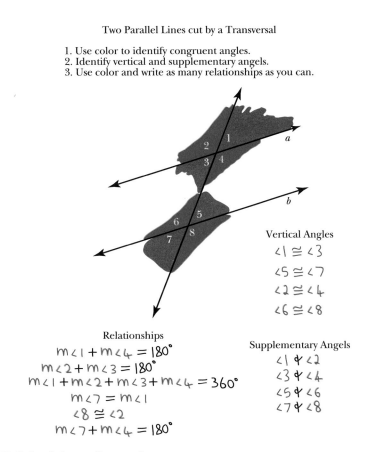

Two Parallel Lines cut by a Transversal

1. Use color to identify congruent angles.
2. Identify vertical and supplementary angels.
3. Use color and write as many relationships as you can.

Vertical Angles

$\angle 1 \cong \angle 3$

$\angle 5 \cong \angle 7$

$\angle 2 \cong \angle 4$

$\angle 6 \cong \angle 8$

Relationships

$m\angle 1 + m\angle 4 = 180°$

$m\angle 2 + m\angle 3 = 180°$

$m\angle 1 + m\angle 2 + m\angle 3 + m\angle 4 = 360°$

$m\angle 7 = m\angle 1$

$\angle 8 \cong \angle 2$

$m\angle 7 + m\angle 4 = 180°$

Supplementary Angels

$\angle 1 \, \& \, \angle 2$

$\angle 3 \, \& \, \angle 4$

$\angle 5 \, \& \, \angle 6$

$\angle 7 \, \& \, \angle 8$

FIGURE 5.21   Color coding angles

heterogeneous students in summer school, we used this strategy a lot to great effect. For example, when one student, Alonzo, finished the staircase task, which asked students to think about pattern growth and the nth case (see Exhibit 5.10), he asked a harder question. He asked how a staircase extending in four directions would grow and the number of cubes in the nth case (see Figure 5.22).

When students are invited to ask a harder question, they often light up, totally engaged by the opportunity to use their own thinking and creativity. This is an easy extension for teachers to use and one that they can have available in any lesson. With any set of mathematics questions, consider giving students a task like this:

"Now you write a question; try to make it hard ☺"

Students can give their questions to other students, who can be encouraged to write questions for each other. This is a particularly good strategy to use for students who work faster than other students or who complain that work is too easy for them, as it involves deep and difficult thinking.

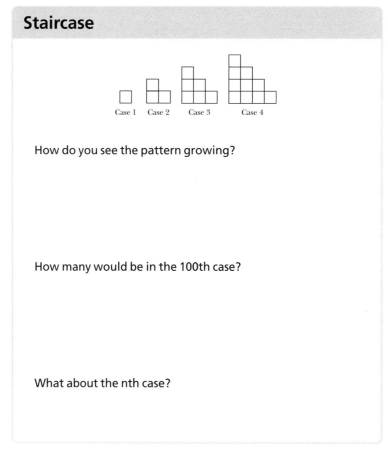

Staircase

Case 1   Case 2   Case 3   Case 4

How do you see the pattern growing?

How many would be in the 100th case?

What about the nth case?

Exhibit 5.10

6. Can You Add the Requirement to Convince and Reason?

Reasoning is at the heart of mathematics. When students offer reasons and critique the reasoning of others, they are being inherently mathematical and preparing for the high-tech world they will be working in, as well as the Common Core. Reasoning also gives students access to understanding. In my four-year study of different schools, we found that reasoning had a particular role to play in the promotion of equity, as it helped to reduce the gap between students who understood and students who were struggling. In every math conversation, students were asked to reason, explaining why they had chosen particular methods and why they made sense. This opened up mathematical pathways and allowed students who had not understood to both gain understanding and ask questions, adding to the understanding of the original student.

FIGURE 5.22 Alonzo's extension problem

I like to accompany one of my favorite tasks for encouraging reasoning with a pedagogical strategy that has many benefits. I learned this strategy from Cathy Humphreys, who asks her students to be skeptics. She explains that there are three levels of being convincing (Boaler & Humphreys, 2005):

Convince yourself

Convince a friend

Convince a skeptic

It is fairly easy to convince yourself or a friend, but you need high levels of reasoning to convince a skeptic. Cathy tells her students that they need to be skeptics, pushing other students to always give full and convincing reasons.

A perfect task to teach and encourage higher levels of reasoning that can be accompanied by the skeptic role was developed by Mark Driscoll; it is called "paper folding." I have used this task with a range of different groups, always with very high levels of engagement. Teachers tell me they love this task, as it often lets students shine who don't typically get that opportunity. In this task, students work in pairs with a square piece of paper. They are asked to fold the paper to make new shapes. Exhibit 5.11 shows the five, progressively more challenging questions (see Figure 5.23).

Paper Folding

Work with a partner. Take turns being the skeptic or the convincer. When you are the convincer, your job is to be convincing! Give reasons for all of your statements. Skeptics must be skeptical! Don't be easily convinced. Require reasons and justifications that make sense to you.

For each of the following problems, one person should make the shape and then be convincing. Your partner is the skeptic. When you move to the next question, switch roles.

Start with a square sheet of paper and make folds to construct a new shape. Then, explain how you know the shape you constructed has the specified area.

1. Construct a square with exactly 1/4 the area of the original square. Convince your partner that it is a square and has 1/4 of the area.

2. Construct a triangle with exactly 1/4 the area of the original square. Convince your partner that it has 1/4 of the area.

3. Construct another triangle, also with 1/4 the area, that is not congruent to the first one you constructed. Convince your partner that it has 1/4 of the area.

4. Construct a square with exactly 1/2 the area of the original square. Convince your partner that it is a square and has 1/2 of the area.

5. Construct another square, also with 1/2 the area, that is oriented differently from the one you constructed in 4. Convince your partner that it has 1/2 of the area.

Source: Adapted from Driscoll, 2007, p.90,
http://heinemann.com/products/E01148.aspx

E x h i b i t   5 . 1 1

When I have given this task to teachers they have struggled for a long time on question 5, some working well into the evening after a full day of professional development, enjoying every moment. Their engagement is enhanced with having a physical shape to consider and change, but also by the need to be convincing. When I give students and teachers this task, I ask for the pairs to take turns, with one folding and convincing and one being the skeptic; then they switch for the

FIGURE 5.23 Teachers work on paper folding task

next question. When I ask students to play the role of being the skeptic, I explain that they need to demand to be fully convinced. Students really enjoy challenging each other for convincing reasons, and this helps them learn mathematical reasoning and proof. As a teacher you may want to model what a fully convincing answer is, by asking students follow-up questions if they have not been convincing enough.

Another example of a task that involves convincing is shown in Exhibit 5.12. The request for students to reason and be convincing can be applied to any mathematics problem or task.

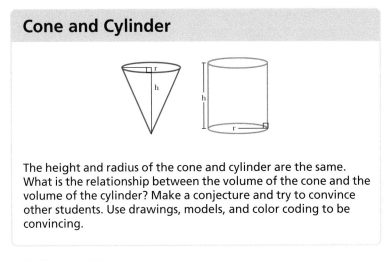

Cone and Cylinder

The height and radius of the cone and cylinder are the same. What is the relationship between the volume of the cone and the volume of the cylinder? Make a conjecture and try to convince other students. Use drawings, models, and color coding to be convincing.

Exhibit 5.12

Conclusion

When mathematics tasks are opened for different ways of seeing, different methods and pathways, and different representations, everything changes.

Questions can move from being closed, fixed mindset math tasks to growth mindset math tasks, with space within them to learn. To summarize, these are my five suggestions that can work to open mathematics tasks and increase their potential for learning:

1. Open up the task so that there are multiple methods, pathways, and representations.

2. Include inquiry opportunities.

3. Ask the problem before teaching the method.

4. Add a visual component and ask students how they see the mathematics.

5. Extend the task to make it lower floor and higher ceiling.

6. Ask students to convince and reason; be skeptical.

Further examples of tasks with these design features are given in Chapter Nine.

If you take the opportunity to modify tasks in these ways, you will be offering your students more and deeper learning opportunities. I have really enjoyed all the times I have seen students working on rich open mathematics tasks, and I have taught with them myself, as students are so excited by them. They love to make connections, which are so important in mathematics, and visual, creative mathematics is inspiring to students. A week of mathematics lessons that include the design features discussed in this chapter, and that are appropriate for grades 3 to 9, can be viewed and downloaded freely here: https://www.youcubed.org/week-of-inspirational-math/.

When I trialed these lessons in middle school classrooms, I experienced parents rushing up to me to tell me that these lessons had changed mathematics for their children. Some parents told me that their children had always disliked math until they took these lessons and saw mathematics in a completely different light. With a design and mathematical mindset, teachers (and parents) can create and transform mathematics tasks, giving all students the rich mathematics environment that they deserve. We cannot wait for publishing companies to realize these changes are needed and make the necessary changes, but teachers can make these changes—creating open, engaging mathematics environments for all of their students.

The following websites provide mathematics tasks that incorporate one or more of the features I have highlighted:

- Youcubed: www.youcubed.org

- NCTM: www.nctm.org (membership required to access some of the resources)

- NCTM Illuminations: http://illuminations.ntcm.org

- Balanced Assessment: http://balancedassessment.concord.org

- Math Forum: www.mathforum.org

- Shell Center: http://map.mathshell.org/materials/index.php

- Dan Meyer's resources: http://blog.mrmeyer.com/

- Geogebra: http://geogebra.org/cms/

- Video Mosaic project: http://videomosaic.org/

- NRich: http://nrich.maths.org/

- Estimation 180: http://www.estimation180.com

- Visual Patterns; grades K–12: http://www.visualpatterns.org

- Number Strings: http://numberstrings.com

- Mathalicious, grades 6–12; real-world lessons for middle and high school: http://www.mathalicious.com

CHAPTER 6

Mathematics and the Path to Equity

I am passionate about equity. I want to live in a world where everyone can learn and enjoy math, and where everyone receives encouragement regardless of the color of their skin, their gender, their income, their sexuality, or any other characteristic. I would like to walk into math classrooms and see all of the students happy and excited to learn, not worrying about whether they look as "smart" as others or whether they have the "math gene." Instead of this vision of equity, mathematics has the greatest and most indefensible differences in achievement and participation for students of different ethnicities, genders, and socioeconomic income levels of any subject taught in the United States (Lee, 2002).

I have been fortunate over many years to conduct research with teachers who have worked in pursuit of equitable math outcomes and who have been extremely successful in achieving them. From this research and from other work with teachers, I have learned the ways in which equitable mathematics classrooms may be promoted, and I will share a range of strategies in this chapter and the next. But first I want to touch on something that is rarely spoken about but that I believe is at the heart of the inequity problem in mathematics.

The Elitist Construction of Math

Mathematics is a beautiful subject, with ideas and connections that can inspire all students. But too often it is taught as a performance subject, the role of which, for many, is to separate students into those with the math gene and those without. Disturbingly, mathematics has been pulled into a culture of performance and elitism in the United States, and I believe that to achieve higher and equitable outcomes we need to recognize the elitist role that mathematics often plays in our society. For mathematics can, on the one hand, be thought of as an incredible lens through which to view the world; an important knowledge, available to all, that promotes empowered young people ready to think quantitatively about their work and lives and that is equitably available to all students through study and hard work. On the other hand, mathematics can be thought of as a subject that separates children into those who can and those who cannot, and that is valuable as a sorting mechanism, allowing people to label some children as smart and others as not smart. Some people revel in the inaccessibility of mathematics as it is currently taught, especially if their own children are succeeding, because they want to keep a clear societal advantage. Others, thankfully, are willing to embrace the change needed, even if their children are succeeding now, especially when they learn that their children's perceived advantage is often based on a math that is really not going to help them in the future.

The Myth of the Mathematically Gifted Child

Some people, including some teachers, have built their identity on the idea they could do well in math because they were special, genetically superior to others. People try really hard to hang on to the idea of children who are genetically gifted in math, and the whole "gifted" movement in the United States is built upon such notions. But we have a great deal of evidence that although people are born with brain differences, such differences are eclipsed by the experiences people have during their lives, as every second presents opportunities for incredible brain growth (Thompson, 2014; Woollett & Maguire, 2011). Even the people whom society thinks of as geniuses actually worked really hard and in exceptional ways to achieve their accomplishments. Einstein did not learn to read until he was nine, and he failed his college entrance examination, but he worked exceptionally hard and had a very positive mindset, celebrating mistakes and persistence. Rather than recognizing and celebrating the nature of exceptional work and persistence, the U.S. education system focuses on "gifted" students who are given different opportunities, not because they show great tenacity and persistence but often because they are fast with math facts. The labeling of students as gifted hurts not only the students who are deemed as having no gifts but also the students who are given the gifted label, as it sets them on a fixed mindset pathway, making them vulnerable and less likely to take risks. When we have gifted programs in schools we tell students that some of the students are genetically different; this message is not only very damaging but also incorrect. Not surprisingly, perhaps, studies that have followed people who had been labeled as gifted in their early years show that they go on to average lives and jobs (http://ireport.cnn.com/docs/DOC-332952).

Malcolm Gladwell unpacks the nature of expertise in his best-selling book *Outliers*. Drawing from extensive research conducted by Anders Ericsson and colleagues, he points out that all experts, including math experts, have worked for at least 10,000 hours in their field (Gladwell, 2011). Some people who have excelled in math choose not to be proud of the hard work and struggle they went through; they prefer to think they were born with a gift. There are many problems with this idea, one being that students who are successful through hard work often think that they are imposters because their achievement was not effortless. Many of these high-achieving students drop out of math because they do not believe that they really belong (Solomon, 2007). This problem comes from a pervasive idea that "math people" are those who effortlessly achieve in math because they were born with something different, and only they truly belong. Add to this idea the stereotyped notions about who is "naturally" good at math, and we start to understand the nature of the problem we face in the United States. Many people recognize that mathematics inequality comes from stereotyped ideas about who can achieve in mathematics and they work to combat them on a daily basis. Unfortunately there are others who work hard, whether consciously or not, to promote the inequities that pervade the mathematics education landscape.

There are some math teachers—fortunately I have met only a few—who think they are superior to teachers of other subjects in their schools, and who think their job is to find the few math students who are special like they are. One high school teacher I met gave 70% of his students an F in every math class he taught, every year. He did not see the students' failure as a reflection on his teaching; he saw it as a reflection on the students who he did not believe had the "gift." In discussions with this teacher, I realized that he feels justified in failing so many students, even though he is ending students' academic futures and stopping them from graduating high school, because he believes he is the guardian of math success and his job is to make sure only the "stars" move on to higher levels. Some university math departments give students a lower grade if they attend office hours and seek help. They do this because the admirable approach of working harder, which should be encouraged, is a sign to them that students don't have the gift. When mathematics is taught with an attitude of elitism, and it is held up as being harder than other subjects and suitable only for the gifted few, a tiny subset of those who could achieve in mathematics—and the scientific subjects which require mathematics—do so. When this elitist idea is combined with stereotypical ideas of who has the gift, harsh inequities are produced. We have only to look at the national U.S. data on the students who take advanced mathematics to see the impact of the elitist, "gifted" culture of mathematics in the United States. In 2013, 73% of math doctorates were male and 94% were white or Asian. The proportion of women taking mathematics PhDs between 2004 and 2013 actually fell, from 34% of students to 27% of students (Vélez, Maxwell, & Rose, 2013). These data should be cause for high-level discussions of mathematics inequities, prompting policy makers and others to seriously consider what we are doing in K–12 schooling that contributes to these growing inequities.

Women are underrepresented in most STEM subjects, but there are also some humanities subjects in which women's underrepresentation is more severe than in STEM. For example, 54% of U.S. PhD students studying the STEM subject of molecular biology are women, but only 31% of students studying the humanities subject of philosophy are women. This was interesting to researchers who looked into the reasons for the different patterns of representation. They found that the subjects in which professors believed that raw, innate talent is the main requirement for success are exactly those subjects in which women—and African American students—are underrepresented (Leslie, Cimpian, Meyer, & Freeland, 2015). As I discussed in Chapter One, math was the STEM subject whose professors were found to hold the most fixed ideas about who could learn. Additionally, researchers found that the more a field values giftedness, the fewer female PhDs there were in the field, and this correlation was found to hold across all 30 fields they investigated. These ideas about giftedness cause fewer women to participate, because strong stereotypes persist about who really belongs in math (Steele, 2011). If women are underrepresented when university mathematics professors believe in giftedness, it is probably safe to assume that the same ideas about giftedness harm girls in early years of schooling, across K-12.

Carol Dweck, Catherine Good, and Aneeta Rattan conducted research to find out how much students felt a sense of belonging in math (Good, Rattan, & Dweck, 2012)—in other words, how much they felt they were members of the mathematics community and how much they felt accepted by those in authority. The researchers found that students' feelings of membership and acceptance in math predicted whether they planned to pursue mathematics in the future. The researchers went on to study the factors in the students' environment that led to different feelings of belonging, and they found that two factors worked against feelings of belonging. One was the message that math ability is a fixed trait; the other was the idea that women have less ability than men. These ideas shaped women's, but not men's, sense of belonging in math. The women's lowered sense of belonging meant that they pursued fewer math courses and received lower grades. Women who received the message that math ability is learned were protected from negative stereotypes—they maintained a high sense of belonging in math and remained intent on pursuing mathematics in the future.

In addition to the ideas of innate talent that pervade mathematics, another problem is the intellectual pedestal upon which most people put mathematics. People who calculate quickly are thought of as smart and special. But why is this? Mathematics is not more difficult than other subjects—I would challenge people who think so to produce a powerful poem or work of art. All subjects extend to difficult levels; the reason so many people think math is the most difficult is the inaccessible way it is often taught. We need to change the thinking around this if we are to open mathematics to many more people.

When Math Inequalities in Course Placement Become Illegal

One source of mathematics inequities is the high school course placement decision-making process. In the United States, the classes that a student takes from ninth grade onward determine, in part, the opportunities they will receive for the rest of their lives. Most universities require at least three years of high school mathematics for college eligibility, making these classes critical for students' futures. This means that high schools should do all that they can to make sure all of their students have the opportunities to take the mathematics courses they need. In my view, because of the role played by mathematics, high school math teachers and their administrators have an extra responsibility to work tirelessly to keep mathematics opportunities open to all students. A recent study of high school placement shed a very interesting—and disturbing—light on this issue.

In 2012 the Noyce Foundation studied student placement in nine school districts in the San Francisco Bay Area and found that over 60% of students who had passed algebra in eighth grade and/or who had met or exceeded state standards on California Standards Tests (CSTs) were placed into an algebra course again when they entered high school, repeating the class they had passed (Lawyers' Committee for Civil Rights of the San Francisco Bay Area, 2013). This started students on a path of low achievement from which many never recovered. In most high schools, only students who start with a class in geometry can ever reach AP statistics or calculus. But why would students be repeating a class, when it was so important for them to start on a higher path and they had already passed an algebra course? When the Noyce Foundation studied the data, they found that the vast majority of the repeating students were Latino/a and African American. The particular data they uncovered showed that 52% of Asian students took Algebra 1 in eighth grade and 52% took geometry in ninth grade. Among white students the algebra participation rate was lower: 59% took algebra in eighth grade but only 33% were in geometry in ninth grade. More disturbingly, 53% of African American students took algebra in eighth grade and only 18% were placed into geometry. Similarly, 50% of Latino/a students took algebra in eighth grade but only 16% were placed into geometry. The filtering of *most* of the African American and Latino/a students who had passed algebra into a low pathway is a clear case of racial discrimination, and the Silicon Valley Community Foundation took the unusual step of hiring lawyers to improve the situation. The legal firm employed found that the schools were acting illegally. They concluded: "Purposeful placement decisions that disproportionately impact minority students violate state and federal laws. But those responsible for math placement decisions also face legal liability if the misplacement decisions are the unintentional results of applying seemingly objective placement criteria that disproportionately impact minority high school students." In other words, math teachers may not be intentionally discriminating by race or ethnicity, but if they use other criteria, such as homework completion, that impact students of color more than other students, they are breaking the law. One of the great achievements of civil rights campaigners in the United States was to make eventual impact the criterion that matters. The San Francisco lawyers highlighted the fact that math placement that results in inequalities is a legal offense.

I expect that the teachers in the study did not think they were blocking the pathways of students of color because of their skin color; rather, a more subtle process of racism was occurring, with teachers deciding that some students do not belong in higher-level mathematics. One middle school principal in a different part of California asked me to sit with him one day and look at his data. He had been disturbed to find that students who had passed algebra in eighth grade in his school were being placed into repeat algebra classes in high school. When we looked together at the data we could not see any relationship between achievement and placement—as we should have—in the data. Rather, we saw a different relationship: a relationship between ethnicity and course placement; the students who were advancing were mainly white, and the students being held back were mainly Latino/a. I immediately recognized this as the same kind of racial discrimination in placement decisions that the Noyce Foundation had revealed. I asked the principal how this could possibly be happening. He explained that the high school teachers had told the middle school teachers they had "better not" advance any student who might fail, and that if students were late with homework or didn't shine in class, they should be held back. One of the high school teachers later managed to instigate a policy across the district that no child with a discipline referral could take algebra in eighth grade in any of the middle schools across the district. Such events may seem incredible, but they are happening because some teachers and administrators believe that they are the guardians of math and it is their job to find the students who truly belong.

After the Noyce Foundation had identified the issue and the Silicon Valley Foundation took on the responsibility of encouraging school districts to improve, a number of school districts made changes in the ways they placed students—with immediate effects. Some of the decisions school districts made were to remove teacher judgment (as sad as it is that this was necessary) and place students in higher-level classes using only course completion and test results. The schools and districts also committed to working quickly over the summer to be able to use test data that are received only weeks before the start of the year, and to appoint a task force to keep watch on the issue and take action to change students' math classes in the early weeks of school if students were placed into lower-level classes than they needed to be in. The racial disparities disappeared almost overnight.

Of course, there is another solution to the problem of students being placed into low-level tracks and classes; that is, to not even offer low-level tracks or classes. We know that when students fail algebra and repeat the course, they typically do the same *or worse* in the repeat class (Fong, Jaquet, & Finkelstein, 2014). This is not surprising, as students repeating a class are delivered a huge message of failure—a message that causes many to decide that they are just not math people and they can never achieve. My preferred solution is to keep expectations and opportunities high for all students and to place all students into geometry or integrated math classes in their first year of high school, whatever their algebra performance, giving them a fresh start. Many people would argue with this, saying that the students have not learned the necessary content to succeed, but geometry offers students opportunities for a new mathematics pathway, and it does not require an algebra course for success. In the next chapter I describe the work of teachers who group students heterogeneously and teach high-level content to all students, showing the impressive impact on student achievement and the strategies they use to bring about success.

In England, students at age 16 take an extremely important final examination in mathematics, the GCSE. The grades students earn on the GCSE will decide what future courses are available to them and even what jobs they can pursue. To enter the teaching profession, for example, candidates need a high grade (A–C) in both the English and the Mathematics GCSE. The mathematics examination is offered at one of two levels. Students entered for the higher paper can achieve any grade from an A* to a D, but students entered for the lower paper can achieve only a C grade or lower. The decision as to which exam paper a student will take is extremely important and is made at a tragically early age for many students, with students being prepared for higher or lower content five or more years prior to the examination. In my study of two schools in England (Boaler, 2002a), one of the schools divided students into high and low sets (like tracks in the United States), and students in the lower groups were prepared for the content in the lower-level exam paper. For three years the students in lower sets were given easier questions, on which they were very successful, and they came to believe they would do well in mathematics. They did not know they were being prepared for a lower-level examination paper, with the highest possible grade only a C.

When students realized they were entered for a lower-level paper, many of them were devastated and simply gave up trying. By contrast, a different school in England took what others thought was a drastic step and entered all of their students into the higher-level paper, whatever their achievement or former preparation. The results were dramatic, with A* to C rates jumping from 40% to over 90%. The head teacher explained to me that they had not made any other changes in the school; they simply started teaching all students higher-level mathematics. The students, receiving such a positive message and opportunity, responded fantastically, stepping up to learn the higher-level content and giving themselves the possibility of a much brighter future. We need all teachers to believe in all students, to reject the idea of some students being suitable for higher-level math and others not, and to work to make higher-level math available to all students, whatever their prior achievement, skin color, or gender. This chapter and the next will talk about ways that teachers can do this.

For mathematics teachers, shifting ideas of who can work on higher-level mathematics is not just a matter for course placement. In their classes, teachers make daily decisions about what students can do that determine the learning paths students will follow. It is natural to plan a math lesson thinking that some students will excel on a task, and to have a pretty good idea of who those students are, but we must all resist such thinking if we are to break the cycle of low achievement that pervades the United States.

I take my undergraduate class each year on a field trip to the incredible Life Academy, a public school in Oakland that is committed to disrupting patterns of inequity on a daily basis. Life Academy is ethnically diverse: 74% of students are Latino/a/Hispanic, 11% African-American, 11% Asian, 2% Filipino, 1% Native American and 1% white, and 92% of students are eligible for free school lunch. The school is located in a part of Oakland in which gang activities and murders are distressingly commonplace. The teachers at Life Academy work hard to make the school a safe space, to communicate to all students that they can achieve at the highest levels, and to inspire them to identify as college-bound students. The mathematics teachers teach heterogeneous math classes using complex instruction, enabling all students to take high-level math classes that they need for college eligibility. The accomplishments of Life Academy are many; the school has the highest college acceptance rate of any high school in Oakland, and the proportion of students who leave "college ready" with California's required classes is an impressive 87%, higher than at the suburban schools in wealthy areas close to Stanford. Some teachers believe that some students cannot achieve at high levels of high school because they live in poverty or because of their previous preparation. In Chapter One I gave an example of high school teachers who made this argument to their school board, but teachers such as those at Life Academy are proving this wrong every day, through teaching high-level mathematics and positive messages to all students.

Recently, in preparation for my online student course, my students and I interviewed a range of passersby in the streets of San Francisco. We interviewed about 30 people, of varying ages, ethnicities, achievement levels, and socio-economic backgrounds. In all the interviews we started with, "Can you tell me how you feel about math?" This showed something really interesting, as every single person immediately told us how well they had performed in math at school. This would not have happened if we had asked them how they felt about art or science or literature. But for people raised in our performance culture, math has played a brutal role as a measurement tool that they have used to judge their self-worth.

I frequently meet parents who are relaxed about their children's learning of English, science, and other school subjects but are extremely anxious about math. Typically such parents want their children to learn high-level math as soon as possible and to take high-level classes as early as possible, as if they will somehow get left behind or lose an advantage if they do not go as fast as possible. This is unfortunate, as we know that students who are advanced in math from an early age are more likely to drop math when they get the opportunity and achieve at lower levels. Bill Jacob is a mathematics professor and vice-chair of the University of California's Academic Senate. When he is asked by districts and parents about accelerating students to higher levels of mathematics, he advises against early acceleration, reporting that a rush to calculus often results in weaker preparation and students dropping courses earlier, which ultimately harms them (Jacob, 2015). He also reports that the BC calculus course does not move students forward mathematically, and students are better off with a stronger preparation in earlier grades. Students who take calculus in high school are highly regarded by universities, but students need only to take calculus by twelfth grade—they do not need to rush through math to take it earlier and produce better-looking transcripts. And calculus is not essential; a number of my Stanford students, even those on STEM pathways, did not take calculus in high school. Recently a parent found her way to my Stanford office to complain because her district had removed advanced classes, so that all students could learn advanced math. She started off aggressively blaming me for the district's decisions, but during the course of our conversation she moved through various emotions, including tears and relief. She first told me her daughter's future had been ruined because she could not take advanced math courses. I explained to her that the district pathway her daughter had been placed on still led to calculus and that she was still learning high-level mathematics in her classes. I also recommended that if her daughter needed more challenge she would benefit more from working through ideas in depth than from getting to higher levels of content faster. The mother calmed down in the course of the conversation and left somewhat reassured but still planning to "home school" her daughter—in math only.

The traditional ways we have taught mathematics and the performance culture that has seeped into the fabric of mathematics teaching and learning harms high-achieving students as much as it does low-achieving students. Research shows us that a high number of high-achieving students drop mathematics, and a decline in conceptual understanding when they are pushed into higher level mathematics classes and tracks (Paek & Foster, 2012). Recently Geoff Smith, the chair of the British and International math Olympiads, spoke publicly about rushing students to higher levels, saying that accelerating students through the system is a "disaster" and a "mistake," and that high-achieving students should explore mathematics in depth rather than rushing onward to higher levels. But there is another way in which the elitist performance culture harms high achievers, which we see in the numbers of students who make the wrong choices for their futures. A study in England showed that undergraduates had chosen mathematics as their university pathway because they had always been good at math. But when they arrived at university they found they were surrounded by other students just as good at math as they were (Solomon, 2007). At that point they experienced a crisis in confidence and in their identities as people (Wenger, 1998). The students had not learned to love math or appreciate the beauty of mathematics; rather, they chose it because they could do it and they had been made to feel they were special. Surrounded by other people apparently as "special" as they were, they lost their purpose and decided to abandon mathematics, realizing they had never developed an interest in the subject itself (Solomon, 2007). For every student that finds herself in a university mathematics pathway when it was not what she really wanted, there are probably a hundred students who could be studying mathematics and enjoying it but have been dissuaded by the faulty image of mathematics given in schools.

Cathy Williams, the executive director of Youcubed, was a district mathematics director for many years before moving to Stanford. As part of her work she met many parents who argued that their students should be given higher levels of content because they were so advanced and so smart. In each case, Cathy would offer to meet with the students, and she would give them an assessment of mathematics to help her understand their needs. Invariably Cathy would find that the students were procedurally fast but could not make sense of mathematics or explain why ideas worked. Students could, for example, divide 1 by 3/4 and get the answer of one and a third, but they could not explain why their answer made sense.

Cathy showed parents that mathematics is a broad subject that goes beyond computation and procedural speed, and involves understanding of ideas. She showed them a visual that highlights three aspects of mathematics (see Figure 6.1).

She then explained to parents that their children were strong in only one area and were only beginning to gain strength in the other important mathematical dimensions. The students did not need more mathematics content as much as they needed to understand the mathematics they had learned, beyond the repetition of procedures, and to be able to apply mathematical ideas. These are the aspects of mathematical thinking that top the list of employers' demands, as Chapter Three showed.

It is not the fault of teachers that an elitist, performance culture has pervaded mathematics, for mathematics teachers are judged by performance scores, just as their students are. The fault lies with our culture, which has favored a role for mathematics as a sorting mechanism and an indicator of who is gifted. There is an imperative need for mathematics to change from an elitist, performance subject used to rank and sort students (and teachers) to an open, learning subject,

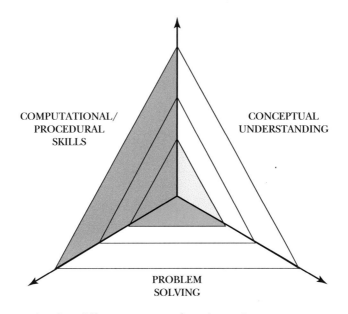

FIGURE 6.1   Balancing different aspects of mathematics

for both high-achieving students, who are currently turning away from mathematics in record numbers, as well as the low-achieving students who are being denied access to ideas that they are fully capable of learning. Many people agree that students need positive mindset beliefs, but if we really want to give these ideas to students, we need to fundamentally change the way mathematics is presented and taught in U.S. society. I end all of my emails to our subscribers to Youcubed with the words "Viva la Revolution!" I do this because it is clear to me that we are in need of a revolution, one that involves changing beliefs about mathematics, the subject, as well as student potential and mindset; one that involves rejecting the elitism that pervades the subject; one that involves moving from performance to learning; and one that involves embracing mathematics as a multidimensional, beautiful subject available to all.

Equitable Strategies

How then do we make math education more equitable? In the next chapters I will talk more about strategies benefitting all students, but here are some strategies for purposefully making math more inclusive.

1. Offer all students high-level content

In the next chapter I will delve into the research and suggested strategies for increasing the numbers of students who are given the opportunity to learn high-level mathematics content. International comparisons have shown that the United States offers fewer students high-level mathematics than do most other countries (McKnight et al., 1987; Schmidt, McKnight, & Raizen, 1997). A clear way of improving achievement and promoting equity is to broaden the number of students who are given high-level opportunities. I devote the next chapter to explaining the best ways to offer high-level mathematics to as many students as possible.

2. Work to change ideas about who can achieve in mathematics

Dweck's studies, as I reviewed earlier in this chapter, show us that the mindset beliefs held by teachers open or close pathways for students, and that fixed mindset thinking and teaching is a large part of the reason inequities continue in mathematics and science, for women and students of color. The studies also show, encouragingly, that students who have a growth mindset are able to shrug off stereotyped messages and continue to success; this speaks again to the huge need for students, and teachers, to develop growth mindset beliefs about their own subjects and transmit growth mindset messages to students. Such messages should be given to students as early and as often as possible, and Chapters One, Two, and Nine review ways to give these messages. It turns out that growth mindset beliefs about learning mathematics may be critical in the pursuit of a more equal society.

3. Encourage students to think deeply about mathematics

In 2014 I was asked to present to the Commission on Women and Girls at the White House. The day was organized around ways to encourage more women to take STEM subjects. I told the gathered group that much of the reason we do not have equal numbers of women and men in STEM is mathematics.

I have found, through my own research (Boaler, 2002b) and other studies that have confirmed the same finding (Zohar & Sela, 2003), that girls, more than boys, desire a depth of understanding that is often unavailable in mathematics classrooms. This is not to say that all girls want something and all boys want something different, but there is a greater tendency among girls to want to understand deeply—to know why methods work, where they come from, and how they connect to broader conceptual domains (Boaler, 2002b). This is a very worthwhile goal and what we want from all of our students. Unfortunately, the procedural nature of mathematics teaching in many classes means that deep understanding is often not available, and when girls cannot gain deep understanding they underachieve, turn away from mathematics, and often develop anxiety. Girls have much higher levels of anxiety about mathematics than boys do (Organisation for Economic Co-operation and Development [OECD], 2015), and the unavailability of deep understanding is one main reason for this (Boaler, 2014a). This is ironic, because the desire to think deeply and really understand concepts is admirable, and the students who express this need are most suited to high-level work in mathematics, science, and engineering. They are the same students who could advance STEM disciplines and break cycles of inequitable teaching. When mathematics is taught procedurally, students who want depth of understanding, most of whom are girls, are denied access to STEM.

In a meta-analysis of 123 informal STEM programs for girls, including summer and after school clubs, researchers summarized the features that girls rated as creating engagement and positive identity formation. The top four features chosen by girls were:

- Hands-on experiences

- Project-based curriculum

- Curriculum with real-life applications

- Opportunities to work together

Role models were also cited, but girls believed them to be less important than opportunities for collaborative, inquiry-based work (GSUSA, 2008). This large-scale study aligns with the research that highlights girls' preferences for a connected approach to mathematics, in which they can pursue questions of why, when, and how methods work. Girls are not alone in preferring this approach, which is also linked to higher levels of achievement, but it seems that girls need this approach more than boys, because without it they are likely to turn away from the discipline.

Learning is not just about accumulating knowledge; it is a process of identity development, as students decide who they are and want to be (Wenger, 1998). For many girls—and boys—the identities they see offered in mathematics and science classrooms are incompatible with the identities they want for themselves (Boaler & Greeno, 2000). Many students see themselves as thinkers and communicators and people who can make a difference in the world (Jones, Howe, & Rua, 2000); in procedural classrooms they often come to the conclusion that they just do not fit in. This relates to the forms of knowledge that are privileged in many mathematics and science classrooms that leave no room for inquiry, connections, or depth of understanding.

When mathematics is taught as a connected, inquiry-based subject, inequities disappear and achievement is increased overall. Chapter Four gave many ideas for teaching mathematics in this way, and Chapter Nine gives many more examples of mathematics tasks, methods, and strategies that enable open, equitable mathematics to be offered to students.

4. Teach students to work together

Many research studies have shown the advantages of students working together for mathematics understanding (Boaler & Staples, 2005; Cohen & Lotan, 2014), and group work is a strategy I regard as critical to good mathematics work (see Figure 6.2). But a fascinating study showed that group work may also be critical in countering racial inequities in mathematics achievement and course taking.

FIGURE 6.2   Students work together in a group

Uri Treisman is a mathematician who worked for many years at the University of California, Berkeley, and is now at the University of Texas. When he was at Berkeley, Treisman was alarmed to find that 60% of African American students were failing calculus, which for many of them meant dropping out of the university. He contrasted the African American students' experiences with that of the Chinese American students, who had much higher rates of success. Treisman studied the reasons for the success differential among the different ethnic groups and found that many of the theories given by professors were not correct—the African American students did not, as some professors thought, have weaker preparation or lower incoming GPAs or come from poorer backgrounds. There was one clear difference between the two cultural groups: The Chinese American students worked on math together. They got together in the evenings after math class, and they worked through problem sets together. When the Chinese American students found mathematics difficult, they were supported—first by knowing that everyone was struggling and then by working together to solve problems. By contrast, the African American students worked on math alone, as a solitary experience, and when they struggled they decided they could not do math. Treisman moved from these results to instigating a new approach at Berkeley in which students were offered collaborative workshops in which they worked through mathematics together and received positive messages about their potential. The impact was dramatic, with failure rates dropping to zero within two years and the

African American students outperforming the Chinese American students who did not attend the seminars (Treisman, 1992).

This is not a solitary finding. Research tells us that when students work on mathematics collaboratively, which also gives them opportunities to see and understand mathematics connections, equitable outcomes result (Boaler & Staples, 2005).

5. Give girls and students of color additional encouragement to learn math and science

Many elementary teachers feel anxious about mathematics, usually because they themselves have been given fixed and stereotyped messages about the subject and their potential. When I taught in my online teacher class that mathematics is a multidimensional subject that everyone can learn, many of the elementary teachers who took it described it as life-changing and approached mathematics differently afterward. Around 85% of elementary teachers in the United States are women, and Beilock, Gunderson, Ramirez, and Levine (2009) found something very interesting and important. The researchers found that the levels of anxiety held by women elementary teachers predicted the achievement of the girls in their classes, but not the boys (Beilock et al., 2009). Girls look up to their female teachers and identify with them at the same time as teachers are often and sadly conveying the idea that math is hard for them or they are just not a "math person." Many teachers try to be comforting and sympathetic about math, telling girls not to worry, that they can do well in other subjects. We now know such messages are extremely damaging. Researchers found that when mothers told their daughters "I was no good at math in school" their daughter's achievement *immediately* went down (Eccles & Jacobs, 1986). Teachers need to replace sympathetic messages such as "Don't worry, math isn't your thing" with positive messages such as "You can do this, I believe in you, math is all about effort and hard work."

In addition to equitable teaching strategies, such as collaboration and inquiry-based approaches, both girls and students of color—particularly underrepresented minorities—need thoughtful and positive messages to be given to them, about their valued place in mathematics. They need this more than other students because of the prevailing stereotyped societal messages about math. The body of work on "stereotype threat," led by the work of Claude Steele, shows clearly the damage caused by stereotypical ideas. Steele and colleagues showed that when girls were given a message that a math test resulted in gender differences, the girls underperformed, whereas girls who did not receive that message performed at the same level as boys on the same test. Steele and colleagues went on to show that a message about gender underachievement did not even have to be given. Subsequent experiments showed that women underachieved when they simply marked their gender in a box before taking the test, compared to those who did not have to do that. He showed, through this and many other studies, that stereotypes are always "in the air" and they reduce opportunities significantly. In subsequent experiments he showed the same impact for white men when playing golf with African American men, as the white men believed they were not as "naturally" good at sports. When they were made aware of race differences before golfing, they performed at lower levels. The work of Steele and colleagues has shown that any group can suffer from stereotype threat when working in an area where another group is believed to be higher achieving (Steele, 2011).

Unfortunately, in math classrooms widespread beliefs about the naturally high achievements of men and White or Asian students are very much "in the air." This makes it critical to address these stereotypical ideas, and one way to do this is by highlighting the achievements of women and underrepresented minorities in mathematics and STEM. In the box I give an example case that could be used in class discussions. There are many more. An ideal way to structure such a class discussion is to ask students to become an expert on the example being discussed, through the jigsaw method that I describe more fully in Chapter Eight, and then share their findings with other students.

Nala Scott, left, 11th grader, and senior Dania Allgood, members of Western High School's all-female robotics team the RoboDoves, with their latest robot, "Joan of Arcs"
Source: Baltimore Sun, used by permission.

A Baltimore all-female, all–African American robotics team has a shelf filled with awards for the remote-controlled robots they have conceived and built.

The RoboDoves team has been so successful they were featured in *Scientific American*. They battle against other high school robotics teams, and the girls show a competitive spirit combined with a love of mathematics, STEM, creativity, and design that could inspire many others. Various news article on this robotics team can be a source of information for students to explore (see Lee, 2014; Zaleski, 2014).

Role models are extremely important to students—and one of the reasons it is so important to diversify the teaching force.

As well as highlighting role models, take other opportunities to encourage students who may need additional encouragement. In my second year of teaching math in a London comprehensive school, I started to celebrate International Women's Day in the school by holding all-girl math sessions in which we worked on interesting math together and celebrated famous female mathematicians. I taught, at the time, at Haverstock School, an inner London secondary school with considerable cultural diversity and over 40 languages spoken by students. One noteworthy outcome of this day of celebration was that it seemed to cause many of the quieter girls, especially those from an Indian background, to gain confidence and participate more in math. Their increased public engagement continued into my math classes afterward.

There are other ways to encourage girls and underrepresented minorities in math. My main point is that it may not be enough, as a math teacher, to treat students equally in the pursuit of equity. Some students face additional barriers and disadvantages, and we must work to address those quite deliberately if we are to achieve a more equitable society.

6. Eliminate (or at least change the nature of) homework

PISA, the international assessment group, with a data set of 13 million students, recently made a major announcement. After studying the relationships among homework, achievement, and equity, they announced that homework perpetuates inequities in education (Program for International Student Assessment [PISA], 2015). Additionally, they questioned whether homework has any academic value at all, as it did not seem to raise achievement for students. This is not an isolated finding; academic research has consistently found homework to either negatively affect or not affect achievement. Baker and LeTendre (2005), for example, compared standardized math scores across different countries and found no positive link between frequency of math homework and students' math achievement. Mikki (2006) found that countries that gave more math homework had lower overall test scores than those that gave less math homework (Mikki, 2006). Kitsantas, Cheema, and Ware (2011) examined 5,000 15- and 16-year-olds across different income levels and ethnic backgrounds and also found that the more time students spent on math homework, the lower their math achievement across all ethnic groups.

It is easy to see why homework increases inequity: students from less-privileged homes rarely have a quiet place to study; they often have to do homework at night, either in the home, while their parents are at work, or at their paid jobs; and they are less likely to have resources such as books and Internet-enabled devices at home. When we assign homework to students, we provide barriers to the students who most need our support. This fact, alone, makes homework indefensible to me.

As a parent who sees her daughters stressed by homework for many nights of the week, with no time for play or the family, I have a personal issue with homework that I want to be very open about. When my eight-year-old daughter said to me last week, "I don't want to do it, I want to sit with you and play," in the two-hour window that is our evening, I did not know what I could say, other than, "Let me write a letter to your teacher and say you are not doing the homework tonight." How reasonable is an eight-year-old's request to spend her evening engaging with her family? My children are in a family of two hard-working parents; we do not even see them until 5:30 at night, when we need to prepare food for them. By the time dinner is finished and we sit down, we have one or two hours before bedtime; this time is rarely spent talking or playing, as homework pressure comes crashing down each night. This is not a good time for my daughters to be encountering difficult problems; often they are simply too tired, so they end up thinking these problems are too hard for them. It is unfair and unwise to give students difficult problems to do when they are tired, sometimes even exhausted, at the end of the day. I wonder if teachers who set homework think that children have afternoon hours to complete it, with a doting parent who does not work on hand. If they do not think this, then I do not understand why they feel they can dictate how children should spend family time in the evenings.

In addition to the inequities created by homework, the stress it causes (Conner, Pope, & Galloway, 2009; Galloway & Pope, 2007), the loss of family time, and the null or negative impact on achievement (PISA, 2015), the quality of math homework is often low, at best. In all of the years in which my eldest daughter was in elementary school, I rarely saw any homework that helped her understanding of mathematics, but I saw a lot of homework that caused her considerable stress. For some reason mathematics teachers—and their textbooks—seem to save the most procedural and uninspiring mathematics for homework. She has been given times tables to memorize; pages of 40 problems to do, all repeating one idea, and lots of questions that she learned to answer correctly in class time and did not need to repeat at home. The value of most math homework across the United States is low, and the harm is significant.

When class starts with a review of homework, inequities are magnified, with some students starting each day behind the other students. When I first moved to the United States, I was shocked to see math classes review homework for 20 to 30 minutes of each lesson. This never happens in England, where homework is treated very differently. In the middle and high schools I know in the United States, every subject gives homework every school night. In England, teachers of different subjects give homework once each week. When I was growing up, I would typically have homework from one subject each night, which took around an hour in the last years of high school. In the United States, at least in my local school district, students in high school regularly stay up until 2 a.m. finishing homework. The levels of stress reported among students are sky high, and one of the main factors causing stress is homework. The significantly smaller amount of homework assigned in the United Kingdom is probably a major reason why homework there gets far less attention and causes far less stress compared with the United States.

If as a teacher or school leader you want to promote equity and take the brave step of eradicating homework, there are many resources that share the research evidence to help you, including Alfie Kohn's *The Case Against Homework,* Sal Khan's arguments in *The One World School House,* and many resources from Challenge Success (for example, Challenge Success, 2012).

If you need to retain homework, then I recommend changing the nature of homework: instead of giving questions students need to answer in a performance orientation, give reflection questions that encourage students to think back on the mathematics of the lesson and focus on the

Exhibit 6.1

big ideas, which we know to be an important orientation for their achievement (PISA, 2012). Exhibits 6.1 and 4.2 both give examples of reflection homework questions.

Alternatively, homework could be an opportunity for giving students inquiry projects; for example, to look for Fibonacci examples in the home and outside. Homework should be given only if the homework task is worthwhile and draws upon the opportunity for reflection or active investigation around the home. If homework was used in this way, and we removed the pages of mindless practice that are sent home daily, we would enable millions of students to use their time more productively, reduce stress, and take a giant step in promoting more equitable schools.

Conclusion

The different equitable strategies I have suggested in the second part of this chapter—changing messages about who belongs; giving more inquiry opportunities; eliminating, reducing, or changing homework; and encouraging group work—are not the usual strategies recommended to teachers when discussions turn to the inequities present in STEM. When I presented recently to the Commission on Women and Girls in the White House, I argued that teaching has often been left out of discussions about the promotion of equity. Organizations worry about role models, and sometimes they are aware of the importance of mindset, but rarely do they consider the huge role played by teaching and teaching methods, which I have highlighted in this chapter. Teachers can make the difference for students who have faced barriers and inequities in their lives. They have the power to do so in the ways they present mathematics and the opportunities they take to encourage vulnerable students. Mathematics is a subject that is critical for all students' futures, as it is a prerequisite for college and many fields. This should mean that mathematics teachers have additional responsibilities—and opportunities—to make mathematics equitably accessible to all. Our society has favored an elitist approach to mathematics, but mathematics teachers—and parents—can reject such messages and open a different pathway for students, one that starts with positive messages about success and the value of persistence and work, and that continues with equitable teaching strategies that empower all students to succeed.

CHAPTER 7

From Tracking to Growth Mindset Grouping

Opportunities to Learn

I still remember clearly the first mathematics lesson I ever taught. It was at Haverstock School in Camden Town, London, a school I described in Chapter Six. When I arrived at the school, the math department used a system of tracking (called "setting" in England) with students working heterogeneously until ninth grade, when they were placed into one of four sets. I walked into my first lesson, excited to teach my ninth-grade students that day, armed with research knowledge on effective teaching. But my students had just been placed into the bottom set. Their first words to me when I greeted them that day were: "What's the point?" I worked hard that year to give them inspirational messages and use the teaching methods I had learned from my training, but their low-level pathway had been mapped out for them, and there was little I could do to change it. In the next year, I worked with the rest of the math department at the school to de-track math classes, and the school has continued to offer all students high-level mathematics ever since.

One key factor in student achievement is known as "opportunity to learn" (OTL). Put simply, if students spend time in classes where they are given access to high-level content, they achieve at higher levels. Of course, this is not surprising to any of us, but what *is* surprising is that even though we know OTL is the most important condition for learning (Wang, 1998; Elmore & Fuhrman, 1995), millions of students are denied opportunities to learn the content they need, and that they could master, as they are placed in low-level classes, sometimes from a very young age. I was shocked by a statistic from England, showing that 88% of students placed into tracks (or sets) at the age of four remained in the same tracks for the rest of their school lives (Dixon, 2002). The fact that children's futures are decided for them by the time they are 4 years old, or even 14 years old, derides the work of teachers and schools and contravenes basic research knowledge about child development and learning. Children develop at different rates and times, and they reveal different interests, strengths, and dispositions at various stages of their development. We cannot know what a 4- or 14-year-old is capable of, and the very best environments we can give to students are those in which they can learn high-level content and in which their interest can be piqued and nurtured, with teachers who are ready to recognize, cultivate, and develop their potential at any time. The new brain science on the incredible capacity of the brain to grow and rewire at any time, along with the evidence on the importance of students' ideas about their own potential, simply adds to the mountain of evidence pointing to the need to move beyond outdated systems of tracking—developed in less knowledgeable times—that continue to limit the achievement of students, no matter what their level of prior achievement.

Once in recent years, as I was presenting to a group of over 800 mathematics teacher leaders, I asked them, "Which current schooling practices deliver fixed mindset messages to students?" Everyone wrote down their top choice, and I collected them. A number of features I have written about in this book, especially assessment and grading (ahead in Chapter Eight), featured prominently, but there was one clear winner: ability grouping. I agree with this assessment. We can give no stronger fixed mindset message to students than we do by putting them into groups determined by their current achievement and teaching them accordingly. The strong messages associated with tracking are harmful to students whether they go into the lowest or highest groups (Boaler, 1997; Boaler, 2013a; Boaler & Wiliam, 2001; Boaler, Wiliam, & Brown, 2001). Carissa Romero, a doctoral student who worked with Carol Dweck and went on to become a director at Stanford, found that the students most negatively hit by the fixed messages they received when moving into tracks were those going into the top track (Romero, 2013).

De-Tracking

In many schools in the United States students are placed into tracked groups for math in seventh grade. When I refer to tracking, I mean the forming of separate classes that provide higher- or lower-level content to students. One important finding of international analysts who study mathematics performance in different countries is that the most successful countries are those that group by ability *the latest and the least*. In the Third International Mathematics and Science Study, for example, the United States was found to have the greatest variability in student achievement—that is, the most tracking. The country with the highest achievement was Korea, which was also the country with the least tracking and the most equal achievement. The United States also had the strongest links between achievement and socioeconomic status, a result that has been attributed to tracking (Beaton & O'Dwyer, 2002). Countries as different as Finland and China top the world in mathematics performance, and both countries reject ability grouping, teaching all students high-level content. San Francisco Unified, one of the larger school districts in California, took the brave step of removing all forms of tracking and all advanced classes before tenth grade. Up until tenth grade all students are encouraged to reach the highest levels they can. All students can still take calculus, and the same high-level classes are available to students in the later years. San Francisco's practice is rare and admirable; the school board unanimously passed the motion to remove earlier forms of tracking after carefully reviewing the research evidence. In most school districts students are filtered into lower and higher pathways at a much younger age. In a school district close to Stanford, in an extremely high-performing community, half of the students are channeled into low-level tracks when they enter seventh grade, which prevents them from eventually taking calculus. It is at that point that parents should hear a strange and unpleasant sound—it is the sound of doors closing on their children's future. If we are to move into a new era in which all students aspire to high levels of mathematics, we need to move to more flexible and research-based forms of grouping, which I will describe later in this chapter.

It is challenging for teachers to give all of their students work that is at an appropriate level for each of them. Teachers know there is a perfect sweet spot—where work is challenging for students but not so challenging as to be out of reach—that creates wonderful classroom engagement. It may seem to make sense that this would be easier to achieve if students are grouped by achievement level. But one reason students perform at lower levels in tracked groups is that students have vastly different needs and backgrounds, even in tracked groups—yet teachers tend to think all the students are the same, and they choose work that is narrow and made up of short questions, which is too easy for some students, too hard for others. This is why the provision of low floor, high ceiling tasks in mathematics classrooms is so important for the future of mathematics in the United States. The other, more obvious reason that tracking reduces achievement is the fixed mindset message that it communicates loudly to all students.

Research has shown what happens when schools and districts decide to de-track. One important study showed the impact of de-tracking in New York City's school district. In New York City, students used to be in middle schools with regular and advanced classes. Then the district decided that they would remove advanced classes and teach advanced mathematics to all middle school students. Researchers were able to follow three years of students working in tracks and then three years when students were working in heterogeneous classes. The researchers followed the six cohorts of students through to the end of high school. They found that the students who worked without advanced classes took more advanced math, enjoyed math more, and passed the state test in New York *a year earlier* than students in tracks. Further, researchers showed that the advantages came across the achievement spectrum for low- and high-achieving students (Burris, Heubert, & Levin, 2006). These findings have been repeated in study after study (see, for example, Boaler, 2013b). A substantial body of research points to the harmful effects of tracking, yet the practice is used in most schools across the country. The rest of this chapter will explain how more modern and effective forms of grouping, that give all students an opportunity and growth mindset messages, can be used in classrooms.

Growth Mindset Grouping

Jill Barshay is a reporter for the *Hechinger Post*. Her popular column "Education by the Numbers" is published weekly in *U.S. News & World Report*. Jill told me that after reading my book *What's Math Got to Do with It* and taking my online teacher course, she became inspired to teach math. She started teaching ninth-grade algebra at a charter school in Brooklyn. But she was not expecting the students she met—a group of demoralized students who had effectively given up on math and themselves because they were not among the students chosen to take algebra in eighth grade. The students told Jill they were not "the smart kids," and they acted out with bad behavior all year. Unfortunately, this is one outcome of tracking students. Most of the unmotivated and bad behavior that happens in classrooms comes from students who do not believe that they can achieve. Teachers worry that de-tracking will cause problems, as badly behaved students will mix with others, but students start behaving badly when they are given the message that they cannot achieve, and who can blame them? In all of my experience teaching heterogeneous groups of students, I have found that when students start to believe they can achieve, and they understand that I believe in them, bad behavior and lack of motivation disappear.

For many years I have worked with one incredible middle school that has a strong commitment to growth mindset teaching and had always grouped students heterogeneously. A few years ago they started experiencing pressure from parents to advance some students so that they could enter high school having completed what has always been a high school class—geometry. Eventually the school gave in to the pressure and started grouping students in regular and advanced math classes. This change was disastrous, resulting in a huge increase in demotivated students across the achievement range. The school reported that students with similar achievement put into different groups experienced huge problems, and many students developed fixed mindsets about their ability. They also found that students who were advanced started to dislike mathematics, and many chose to drop out of the advanced class, which damaged them further. Within two years the school abandoned the tracking system and put students back into heterogeneous groups. Now they offer geometry as an optional class before school for anyone who wants to take it. This is an excellent strategy to deal with the pressure from parents, as it allows an option to students who want to take more advanced classes without communicating harmful fixed mindset messages to all students about their potential. It was very important that geometry was offered as a choice to every student at the school.

Teachers who want to offer high-level opportunities to all students but are forced to teach with tracks may choose to teach all students high-level work, whatever track they are in. The teachers I have worked with who have done this know that the tracking is limiting students' achievement and that students in the lower-level classes they are given are capable of the higher-level work, when given the right messages and teaching.

In another excellent middle school in an urban area, which was committed to growth mindset teaching, the teachers de-tracked and offered the choice of a class intended to support lower-achieving students. Again, the class was offered to everybody, for any students who wanted some more time to go into depth. The extra class followed the regular math class in the time table and it did not work on remediation; rather, it was an opportunity to revisit the math from the regular class and talk about it, returning to the class ideas and going into some more depth. Many students chose to take the extra class—all of the students who were finding math difficult, as well as those who were not but wanted more depth. It's important to note that the class was a choice for any student, and the class title did not suggest that it was for low achievers.

Teachers who are committed to a different future—one in which all students have growth mindsets and opportunities—and who choose to teach heterogeneous classes, are admirable. But the teaching of groups of students with a wider range of previous achievement requires knowledgeable teaching. It is not enough to de-track and then teach through narrow mathematics questions that will be accessible to only a few students. Over the years I have been fortunate to work with many incredible teachers, committed to equity, who teach mixed-achievement groups with great success. In the remainder of this chapter I will share some important strategies for teaching heterogeneous groups effectively—strategies supported by research evidence.

Teaching Heterogeneous Groups Effectively: The Mathematics Tasks

When de-tracking math classes, it is very important that we offer students the opportunity to take mathematics to different levels and not give them closed math questions suitable for only a small subset of the class. There are different ways in which students can be encouraged to take mathematics to different levels.

1. Providing Open-Ended Tasks

As I explained in Chapter Five, if closed questions are given to students in heterogeneous groups, many will fail or not be challenged, so it is imperative that tasks are open-ended, with a low floor and a high ceiling (see Figure 7.1). Low floor, high ceiling tasks allow all students to access ideas and take them to very high levels. Fortunately, low floor, high ceiling tasks are also the most engaging and interesting math tasks, with value beyond the fact that they work for students of different prior achievement levels. They are tasks that teach important mathematics, inspire interest, and encourage creativity. Chapter Five gave a range of examples of such tasks and links to websites that provide them.

At Phoenix Park, a highly successful school in England that used project-based methods, the teachers had collected a range of low floor, high ceiling tasks that students could take to any level. Some students would take the tasks to very high levels on some days, other students on other days. It was impossible, as it should be, to predict which students would take math to higher levels on any given day. In Chapter Five I gave the example of the "maximum area enclosed by a fence" question; for some students this meant that they learned about trigonometry, for others it meant learning about Pythagoras, for still others it meant learning about shape and area. The teachers' role in the classrooms was to discuss the mathematics students were working on, to guide them, and to extend their thinking. In a traditional classroom the textbook is meant to do this, through the mathematics it presents, but the textbook is a very blunt instrument that cannot tell what a student knows or needs to know. In a *growth mindset* classroom the teacher is the one making these decisions in relation to individuals or groups of students, to challenge, support, and stretch them at exactly the right level. The opportunities that teachers have to interact with students as they work on open tasks and to introduce them to mathematics, and hold important discussions with them, is one reason students do so well in such teaching environments. Such teaching, although demanding, is also extremely fulfilling for teachers, especially when they see students who lack confidence and were previously low-achieving take off and soar.

A few years ago in England I worked with a group of teachers who had decided to de-track their middle and high school classes after learning about the "complex instruction" method I describe

The Pocket Game

Circles and Stars

Ice Cream Scoop

FIGURE 7.1   Open-ended tasks from Youcubed

next. The teachers had not had any special training and they had not developed the wonderful curriculum used at Phoenix Park, but they had learned about complex instruction and had collected some low floor, high ceiling tasks. At the end of the first week of teaching the new classes, grouped for a growth mindset, one teacher exclaimed in amazement that after he gave out one task, a student who "would have been in the bottom group" was the first to solve it. Over time the teachers continued to be surprised and pleased by the different creative methods shown by different students from across the achievement range. The teachers were thrilled with how well students responded to the de-tracking and with how disciplinary issues, which they had feared would increase, disappeared almost overnight. This was interesting to me, as the teachers had been quite worried about de-tracking and whether the students would work well together. They discovered that when they gave open tasks, all students were interested, challenged, and supported. Over time the students they thought of as low-achieving started working at higher levels, and the classroom was *not* divided into students who could and students who could not; it was a place full of excited students learning together and helping each other.

2. Offering a Choice of Tasks

Students in growth mindset classrooms do not always need to work on the same tasks; they can be offered different tasks that address different levels and fields of mathematics. What is important is that students are able to choose the task they want to work on instead of the choices being made by teachers. On one occasion when I was observing at Phoenix Park School, the students were offered a choice between two tasks: (1) to investigate shapes with an area of 64 or (2) to investigate shapes with a volume of 216. In a fourth-grade classroom I watched a teacher ask the students to use fraction strips or Cuisenaire rods to find as many fractions as they could that were equivalent to 1/4, and as an extra challenge to look for fractions equivalent to 2/3. This provision of extension activities and different tasks that have extra challenge is something that can (and probably should) be done in every lesson. The choice or the challenge must always be available to all students.

Some students, at times, may need a push to take the higher-level task that is offered. What is important when tasks are on offer is that students never get the idea that they can work on only the low-level task or that teachers don't think they are capable of the higher-level task. When I have seen this strategy in use with different teachers, they have communicated to students that the tasks go to different places, or that some provide an extra level of challenge, and students are happy to be given the chance to consider what they want to work on and to have the excitement of being offered an extra challenge.

3. Individualized Pathways

When I was teaching the heterogeneous groups at Haverstock School, I used a mathematics series that had been designed for use in mixed-achievement groups in urban settings. The scheme is called SMILE, which stands for Secondary Mathematics Individualised Learning Experience. London is an amazing and diverse city with a high rate of student turnover. Teachers in inner

FIGURE 7.2   SMILE cards

London know that they may have one set of students to teach one day and a changed class the next, with some students absent or having left the class and others joining the class. Many teachers in central London were committed to mixed achievement and to heterogeneous grouping. Unusually, SMILE was a set of mathematics "cards" (actually full-sized sheets) designed by teachers to be engaging and culturally sensitive (see Figure 7.2). Thousands of cards were written, with each card teaching a different part of mathematics. Any inner London teacher could submit a new card, and over time the collection grew to over 3,000 interesting cards, all written by teachers. SMILE teachers would work by assigning each student 10 cards, which the students would work through and show to the teacher, who would then give 10 more cards. The cards were individually assigned, but many asked students to find a partner to work on the mathematical ideas together.

As the cards were individualized, students could work through them at their own pace, and teachers would walk around and help the students. In my own experience of teaching SMILE, I observed highly engaged students who would excitedly collect their cards, knowing that their achievement was in their own hands. On some days we would work not with the cards but as a whole class on a mathematics investigation. SMILE was highly effective in an urban mixed-ability setting, as it allowed individualized work, and student absences did not cause problems in the flow of the class. Many of the SMILE cards (available from http://www.nationalstemcentre.org .uk/elibrary/collection/44/smile-cards) are excellent, although they may need adapting for other locations, as they are written for students in London with many London examples.

The advent of technology has meant that individualized mathematics have become more widely available. Sal Khan, the founder of Khan Academy, is a famous advocate of individualized learning. He correctly highlights the crudeness of tracking and shows how students who are allowed to choose their content and learning path can achieve at incredibly high levels, from any starting point (Khan, 2012). Other technology-based companies have produced products that allow students to work at their own levels. Sadly, however, I have yet to encounter a product that gives individualized opportunities and also teaches mathematics well. But the principle of allowing students to form their own learning pathways and encounter individualized content, when accompanied with opportunities for group work and collaboration, offer potential for high-level work to be offered to all students.

Teaching Heterogeneous Groups Effectively: Complex Instruction

The mathematics tasks that are used in heterogeneous classrooms are extremely important, but so are the norms and expectations that are set up for the ways students work together. Experienced teachers know that group work in classrooms can fail when students participate unequally in groups. If students are left to their own devices and they are not encouraged to develop productive norms, this is fairly likely to happen: some students will do most of the work, some will sit back and relax, some may be left out of the work because they do not have social status with other students. Elizabeth Cohen, a sociologist at Stanford, observed unequal group work patterns in classrooms and realized that it was due to social differences in groups, with some students being assigned or assuming a status of importance while others were given a low status label (Cohen, 1994). My Stanford colleague Jennifer Langer-Osuna has studied many cases of group work in which the perceived status of the student speaking is the reason ideas are taken up, rather than the mathematics within the ideas (Engle, Langer-Osuna, & McKinney de Royston, 2014). She has also found that status differences often come from stereotyped beliefs about students of a certain race, class, or gender (Esmonde & Langer-Osuna, 2013; Langer-Osuna, 2011). Liz Cohen, with Rachel Lotan, went on to design "complex instruction," a pedagogical approach, designed to make group work equal, that can be used with any grade level or school subject (Cohen & Lotan, 2014).

In a four-year National Science Foundation (NSF) research study, I contrasted different approaches to mathematics teaching. My team of graduate students and I followed over 700 students through four years in different high schools (Boaler, 2008; Boaler & Staples, 2005). Approximately half of the students were in schools that worked in tracked groups, through procedural math teaching and testing. The other half were at an urban high school in California, which I called Railside, in which the teachers de-tracked classes and taught mathematics using complex instruction. The students at Railside were racially diverse, with more English language learners and higher levels of cultural diversity than the other schools in our study. At Railside approximately 38% of students were Latino/a, 23% African American, 20% White, 16% Asian or Pacific Islanders, and 3% from other groups. In the schools teaching mathematics traditionally, 75% were white and 25% were Latino/a. At the start of our study, when students had just completed middle school, we administered an assessment of middle school math. At that time the students at Railside were achieving at significantly lower levels than the students in the other, suburban schools in our study, which is not atypical in urban settings where students have many issues to deal with in their lives (see Figure 7.3).

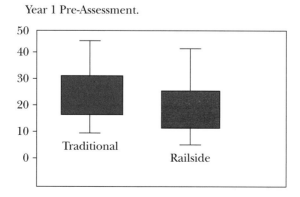

FIGURE 7.3   Pre-assessment test scores

Year 1 Pre-Assessment.

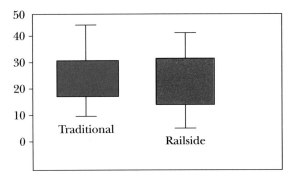

FIGURE 7.4   Year 1 Assessment test scores

Year 2 Pre-Assessment.

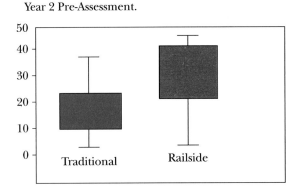

FIGURE 7.5   Year 2 Assessment test scores

After one year the Railside students had caught up with the students working traditionally (Figure 7.4).

Within two years they were achieving at significantly higher levels (Figure 7.5).

In addition to their higher performance, the students at Railside enjoyed math more and continued to higher levels. At Railside, 41% of students took advanced classes of pre-calculus and calculus, compared with only 27% of the students who learned traditionally. In addition, all racial inequities in achievement decreased or disappeared while the students were at Railside School (Youcubed at Stanford University, 2015a; https://www.youcubed.org/category /making-group-work-equal/).

Recently an important book devoted to understanding Railside and all of its equitable practices has been produced, written by researchers and teachers from Railside School (Nasir, Cabana, Shreve, Woodbury, & Louie, 2014).

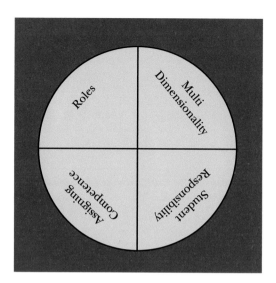

FIGURE 7.6 Complex instruction

For the remainder of the chapter, I will consider how the school brought about these impressive achievements, under the four tenets of complex instruction: multidimensionality, roles, assigning competence, and shared student responsibility (see Figure 7.6).

Multidimensionality

A one-dimensional math class, of which there are many in the United States, is one in which one practice is valued above all others—usually that of executing procedures correctly. This narrow criterion for success means that some students rise to the top of one-dimensional classes, gaining good grades and teacher praise, while others sink to the bottom, with most students knowing where they are in the hierarchy created. Such classrooms are one-dimensional, because in them there is only one way to be successful. In a multidimensional math class, teachers think of *all* the ways to be mathematical. If we consider the work of mathematicians, for example, we know that they perform calculations at some times, but they also have to ask good questions, propose ideas, connect different methods, use many different representations, reason through different pathways, and many other mathematical acts. Mathematics is a broad and multidimensional subject. In complex instruction (CI) classrooms, teachers value, and assess students on, the many different dimensions of math. The mantra of the CI approach, which was posted on the classroom walls at Railside, is:

> No one is good at all of these ways of working, but everyone is good at some of them.

When we interviewed students in our study, we asked: "What does it take to be successful in math?" A stunning 97% of students from the traditional approach said the same thing: "Pay careful attention." This is a passive learning act that is associated with low achievement (Bransford,

Brown, & Cocking, 1999). In the Railside classes, when we asked students the same question, they came up with a range of ways of working, such as:

- Asking good questions

- Rephrasing problems

- Explaining

- Using logic

- Justifying methods

- Using manipulatives

- Connecting ideas

- Helping others

A student named Rico said in an interview, "Back in middle school the only thing you worked on was your math skills. But here you work socially and you also try to learn to help people and get help. Like you improve on your social skills, math skills, and logic skills" (Railside student, year 1).

Rico chose to talk to us about the breadth of the mathematics he was experiencing. Another student, Jasmine, added, "With math you have to interact with everybody and talk to them and answer their questions. You can't be just like 'Oh here's the book, look at the numbers and figure it out.'" When we asked, "Why is that different for math?" she said, "It's not just one way to do it. It's more interpretive. It's not just one answer. There's more than one way to get it. And then it's like: 'Why does it work?'" (Railside student, year 1). Jasmine highlights the social and "interpretive" nature of mathematics, with students going beyond the book and the numbers to make sense of mathematical ideas, consider different approaches, and justify their thinking, answering the important question, "Why does it work?"

At Railside the teachers created multidimensional classes by valuing many dimensions of mathematical work. This was achieved by giving rich tasks that the teachers described as group-worthy problems—problems that were difficult to solve alone and that required different members of a group to contribute. Lani Horn describes group-worthy tasks as those that "illustrate important mathematical concepts, allow for multiple representations, include tasks that draw effectively on the collective resources of a group, and have several possible solution paths" (Horn, 2005, p. 22). Two examples of questions that would be deemed group worthy, which come from nrich.maths.org, are in Exhibits 7.1 and 7.2. The full task, with handouts, can be found in the appendix.

The third example (Exhibit 7.3) comes from Railside, where teachers first gave students linear function tasks (they called them pile patterns) that showed a particular representation and asked students to predict, for example, pile number 10.

Some teams might answer the question geometrically; some numerically, with a t-table; some algebraically. After asking students to share their solutions, teachers would ask "Did anyone see it in a different way?"

Sorting the Numbers

Well, how about doing a simple jigsaw?

This problem has been designed to be worked on in a group of about four. (Some teacher notes and ideas for an extension are at http://nrich.maths.org/6947&part=note.)

1. There are two jigsaw puzzles that your teacher can print out for you (see below).

Complete each jigsaw and then put the pieces into the outline squares, which can be printed:

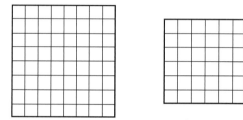

2. Place the smaller square of numbers on top of the other larger square in any way you like so that the small centimeter squares match up. (You may find it easier to copy the numbers on the smaller square onto a transparent sheet.)

3. Explore what happens when you add together the numbers that appear one on top of the other.

4. In your group, explore any other ideas that you come up with.

Exhibit 7.1

Later, Railside teachers moved on to giving students challenge activities that did not give all the information the students needed, so they had to work together to generate missing elements of the table, graph, equation, and geometric representations of the pattern, as shown in Figure 7.7.

More detail is given on these and other tasks used at Railside in Nasir et al. (2014), and many of the tasks are available in the CPM Connections series. The teachers at Railside had once taught tracked groups using traditional methods, but there was a high percentage of math failure among the students. The teachers at Railside did not assume the math failure was due to the

When you've looked at the 36 combinations, you probably need to ask, "I wonder what would happen if we …?" Change one small thing, explore that, and then compare your two sets of results. You might like to ask, "Why…?"

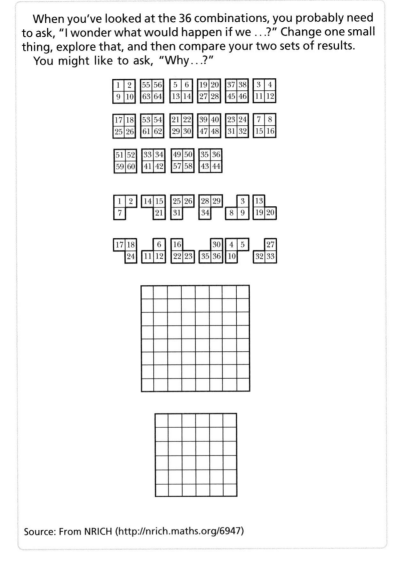

Source: From NRICH (http://nrich.maths.org/6947)

E x h i b i t   7 . 1   (Continued)

students' weakness, even though many students arrived at the school with the equivalent of only second-grade mathematics knowledge. Instead, the teachers applied for a grant that enabled them to spend a summer planning a new curriculum and approach. The teachers had learned about complex instruction, so they de-tracked classes and designed an introductory algebra curriculum that all incoming high schoolers would take. They designed the algebra course to have considerably more depth than traditional courses, to provide a challenging experience for all students even

Growing Rectangles

Imagine a rectangle with an area of 20cm².

What could its length and width be? List at least five *different* combinations.

Imagine enlarging each of your rectangles by a scale factor of 2:

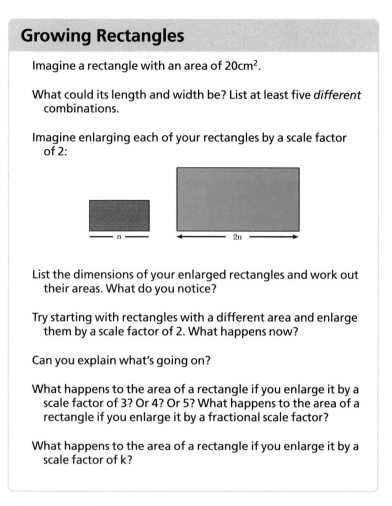

List the dimensions of your enlarged rectangles and work out their areas. What do you notice?

Try starting with rectangles with a different area and enlarge them by a scale factor of 2. What happens now?

Can you explain what's going on?

What happens to the area of a rectangle if you enlarge it by a scale factor of 3? Or 4? Or 5? What happens to the area of a rectangle if you enlarge it by a fractional scale factor?

What happens to the area of a rectangle if you enlarge it by a scale factor of k?

E x h i b i t   7 . 2

if they had taken an algebra class before. As the teachers at Railside were deeply committed to equity and to heterogeneous teaching, they worked together to develop and implement a curriculum that gave multiple access points to mathematics. Standard textbooks are usually organized around mathematical methods such as "graphing linear functions" or "factoring polynomials." The Railside teachers organized their curriculum around big ideas such as "What is a linear function?" The teachers did not design tasks; instead, they chose deep, conceptual math tasks from different published curricula, such as the College Preparatory Mathematics curriculum (CPM) and the Interactive Mathematics Program (IMP), and they chose to represent algebra not only visually but also physically, building the curriculum around algebra lab gear, a physical manipulative used to develop algebraic understanding, shown in Figure 7.8 (Picciotto, 1995).

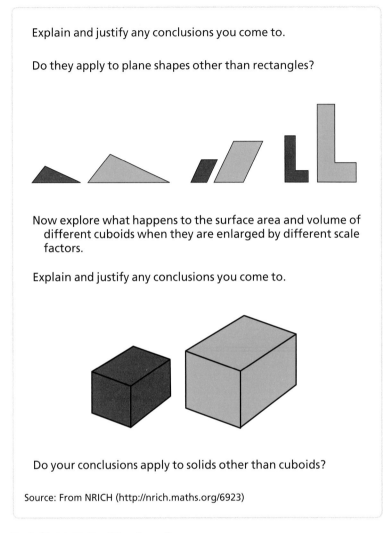

Explain and justify any conclusions you come to.

Do they apply to plane shapes other than rectangles?

Now explore what happens to the surface area and volume of different cuboids when they are enlarged by different scale factors.

Explain and justify any conclusions you come to.

Do your conclusions apply to solids other than cuboids?

Source: From NRICH (http://nrich.maths.org/6923)

Exhibit 7.2 (Continued)

A theme of the algebra course, and then later all the courses in the school, was multiple representations—students were frequently asked to represent their ideas in different ways, such as through words, graphs, tables, symbols, and diagrams. Students were also encouraged to color code, representing ideas in the same color—for example, using the same color for the x in an expression, diagram, graph, table, and paragraph (see Exhibit 7.4).

The multidimensional nature of the classes at Railside was an extremely important factor in the students' increased success. When we analyzed the reasons for the students' widespread high achievement at Railside, we realized that, in essence, many more students were successful because there were many more ways to be successful.

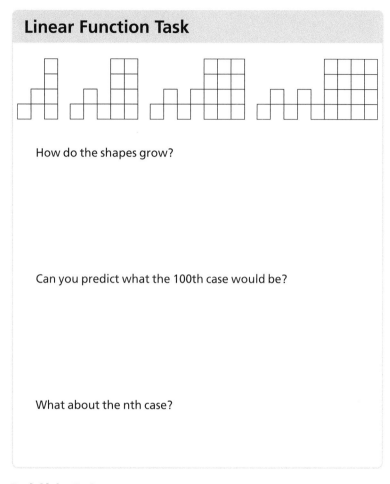

Linear Function Task

How do the shapes grow?

Can you predict what the 100th case would be?

What about the nth case?

Exhibit 7.3

The Railside teachers valued many dimensions of mathematical work and also used multidimensional grading (see Chapter Eight). Although the standardized state tests the students had to take under the State of California requirements did not value multidimensional mathematics, the students achieved at high levels because they had learned to be successful in class and to feel good about mathematics. They also approached the state tests as confident problem solvers willing to try any question. The Railside students' state test scores were higher in mathematics than all other subjects in the school—which is very unusual—and the school outperformed all other schools in their district in math, despite being in the lowest-income area.

In one algebra class I observed, the students had been given a typically challenging problem with few instructions. The students were asked to use their math tools, such as t-tables and graphs, to produce an equation in the form y = mx + b, that would help them know the length of shoelaces they needed to buy for different shoes (see Exhibit 7.5).

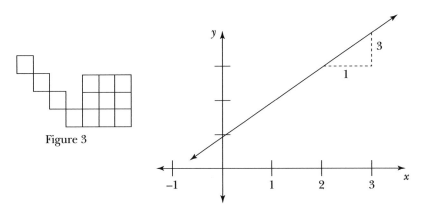

Figure 3

FIGURE 7.7 CPM task

FIGURE 7.8 Students work on finding the perimeter of a shape built from algebra lab gear

The teacher encouraged the groups to work with a real shoe, contributed by a group member. She introduced the problem by telling students that there were lots of ways to start the problem and that success on the problem would take good communication among team members, with students listening to each other and giving each other a chance to think through their work. The teacher also explained that students would get a better grade on the problem if they used multiple methods to show and explain their work.

As with many mathematics questions, the most difficult aspect of the problem for many of the students was the beginning—knowing how to start. They had been told to form an equation to

Math Function Task

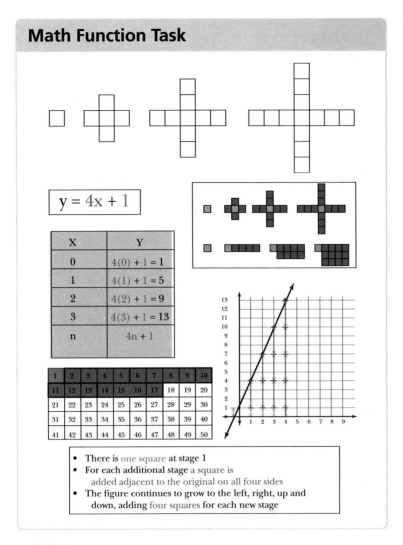

- There is one square at stage 1
- For each additional stage a square is
  added adjacent to the original on all four sides
- The figure continues to grow to the left, right, up and
  down, adding four squares for each new stage

Exhibit 7.4

help them buy shoelaces—a fairly open instruction, leaving students to work out that certain variables, such as the number of lace holes and the length of laces needed to tie a bow, could be represented in their equation. They also needed to work out that y in the equation needed to represent the length of shoelaces needed.

As I watched the class work, I noticed that many of the groups did not know how to start the problem. In one group a boy quickly announced "I don't get it" to his group, and one of his group members agreed, saying, "I don't understand the question." At that point a girl in the group suggested that they reread the question out loud. As they read it, one boy asked the others, "How

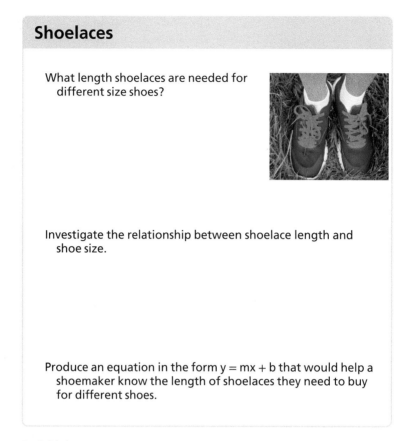

Shoelaces

What length shoelaces are needed for different size shoes?

Investigate the relationship between shoelace length and shoe size.

Produce an equation in the form y = mx + b that would help a shoemaker know the length of shoelaces they need to buy for different shoes.

Exhibit 7.5

is this shoe connected to that equation?" Another boy suggested that they work out the length of their own shoelace. The group set to work measuring the lace, at which point one boy said that they would need to take into account the number of lace holes needed (see Figure 7.9). The group continued, with different students supporting others by asking questions for the group to consider.

I watched many examples like this one: students were able to get started through encouraging each other, rereading questions, and asking each other questions. The students had been encouraged to read problems out loud, and when they were stuck, to ask each other questions such as:

- What is the question asking us?

- How could we rephrase this question?

- What are the key parts of the problem?

Part of the Railside teachers' approach was to give groups a problem and, when each group had finished, to ask a follow-up question to assess students' understanding. Through the questions teachers asked the students and the encouragement they gave, such as asking students to restate

FIGURE 7.9   Students work to produce an equation representing the length of shoelaces needed for different shoes

a problem, the students became very skilled at asking similarly helpful questions of each other. Before long, high levels of engagement spread through the class as the students began measuring laces and thinking about relationships with lace holes. The students' engagement was due to many factors:

- The work of the teacher, who had carefully set up the problem and circulated around the room asking students questions

- The task itself, which was sufficiently open and challenging to allow different students to contribute ideas

- The multidimensionality of the classroom: different ways to work mathematically, such as asking questions, drawing diagrams, and making conjectures were valued and encouraged

- The request to deal with a real-world object and idea

- The high levels of communication among students who had learned to support each other by asking each other questions.

Many mathematics departments employ group work, but they do not experience the high rates of success among students and the impressive work rate that we witnessed in groups at

Railside. Part of the reason students worked so well at Railside was that multidimensional mathematics was taught and valued, and the teachers taught students to support each other's learning.

Roles

When students were placed into groups, they were each given a role to play. Exhibit 7.6 shows one of the task sheets given to students, with roles explained.

Group Roles USA
Facilitator:

Make sure your group reads all the way through this card together before you begin. "Who wants to read? Does every one get what to do?"

Keep your group together. Make sure everyone's ideas are heard. "Did anyone see it a different way? Are we ready to move on?" Be sure everyone can explain.

Recorder/Reporter:

Your group needs to organize all your results. Your results need to show everyone's ideas, be well organized, and use color, arrows, and other math tools to communicate your mathematics, reasons, and connections. "How do we want to show that idea?" Be ready to join the teacher for a huddle.

Resource Manager:

- Get materials for your team.
- Make sure all questions are team questions.
- When your team is done, call the teacher over to debrief the mathematics.

Team Captain:

- Remind your team to find reasons for each mathematical statement and search for connections among the different statements. "How do you know that for sure? How does that relate to …?"
- No talking outside your group!

Exhibit 7.6

Group Roles, British

Organiser:

- Keep the group together and focused on the problem; make sure no one is talking to people outside the group.

Resourcer:

- You are the only person that can leave their seat to collect rulers, calculators, pencils, etc., for the group.
- Make sure everyone is ready before you call the teacher.

Understander:

- Make sure all ideas are explained so everyone is happy with them.
- If you don't understand, ask whoever had the idea … if you do, make sure that everyone else does too.
- Make sure that all the important parts of your explanation get written down.

Includer:

- Make sure everyone's ideas are listened to; invite other people to make suggestions.

Exhibit 7.7

When I introduced the complex instruction approach to teachers in England they changed some of the roles, wanting the roles to sound more British and show less hierarchy. The teachers in England decided on the names and descriptions that they gave students, shown in Exhibit 7.7.

The roles are an important part of complex instruction because they give everybody a part to play and they encourage student responsibility. As the UK example shows, the roles can be adapted by teachers for their particular classroom needs.

In complex instruction classrooms, a chart of groups and roles is placed on the wall. Students are randomly assigned to groups and roles (see Figure 7.10).

Every few weeks, students are asked to change groups and given different roles. The teachers emphasize the different roles at frequent intervals during class, stopping, for example, at the start of class to remind facilitators to help people check answers or show their work or to ask questions.

All teachers know that it can be difficult to bring students back to listen to them after students have started work in groups and are excitedly talking about mathematics. But teachers frequently need to inject a new piece of information or a new direction into the group work. In complex instruction, teachers do not try to do this by quieting the whole class. Instead, they call the recorder/reporters out to join the teacher for a huddle. The recorder/reporters meet as a

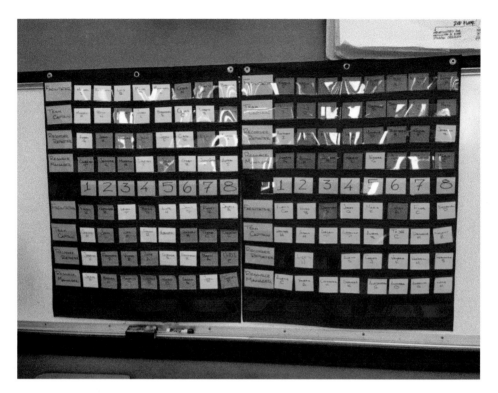

FIGURE 7.10   Random assignment of groups and roles

group with the teacher, who can give information that each recorder/reporter takes back to the groups. This not only helps the teacher but also gives the students responsibility that is intrinsically valuable in helping them feel empowered mathematically. The complex instruction roles contributed to the interconnected system that operated in Railside classrooms, a system in which everyone had something important to do and the students learned to rely upon each other.

Assigning Competence

An interesting and subtle approach recommended in the complex instruction pedagogy is that of *assigning competence*. This practice involves teachers' raising the status of students who they think may be lower status in a group—by, for example, praising something they have said or done that has intellectual value, and bringing it to the group's or the whole class's attention. For example, teachers may ask a student to present an idea or publicly praise a student's work in a whole class setting.

I did not fully understand the practice of assigning competence until I saw it in action. I was visiting Railside one day and watching a group of three students work. As the teacher came over, a quiet Eastern European boy muttered something to the other two members of his group—two happy and excited Latina girls who were directing the flow of the work. Ivan said something really brief: "This problem is like the last problem we did." The teacher who was visiting the table immediately picked up on it, saying "Good, Ivan, that is a really important point; this is like the last problem, and that is something we should all think about—what is similar and different about the two problems." Later, when the girls offered a response to one of the teacher's questions, he said, "Oh, that is like Ivan's idea; you're building on that." The teacher raised the status of Ivan's contribution, which almost certainly would have been lost without his intervention. Ivan visibly straightened up and leaned forward as the teacher praised his idea and later reminded the girls of his idea. Cohen (1994), the designer of complex instruction, recommends that if student feedback is to address status issues, it must be public, intellectual, specific, and relevant to the group task (p. 132). The public dimension is important, as other students learn that the student offered the idea; the intellectual dimension ensures that the feedback is an aspect of mathematical work; and the specific dimension means that students know exactly what the teacher is praising.

Teaching Students to Be Responsible for Each Other's Learning

A major part of the equitable results achieved at Railside and a central part of the CI approach is teaching students to be responsible for each other's learning. Many schools use group work and hope that students will develop responsibility for each other, but this continues to be a challenge. At Railside, the teachers used many different methods to help students learn to work well together and develop responsibility for each other. One important decision the teachers made was to spend 10 weeks in the first algebra class at the start of the first year, teaching students how to work well in groups. The students worked on math during this time, but the math was the teachers' secondary goal; their main priority was teaching students how to work well together. Anybody visiting Railside's classes after that 10-week period would see how that time investment paid off, with students engaging in respectful conversation, listening to each other, and building upon each other's ideas.

When I teach students in groups, I also first spend time developing careful group norms of respect and listening. An activity I always like groups to work on before introducing any math work is to ask them to discuss together the things they do and don't like other group members to do and say when they are working in a group on math. We make two big posters together—one, of the things people like, and the other, of the things they don't like—to display on the walls (see Figure 7.11). In the "don't like" list, students usually raise many of the actions that I also like to discourage, such as one person doing the work and then telling everyone the answer, or people being patronizing and saying things like "This is easy," or leaving people out of discussions. I find that when students think for themselves about positive and negative group discussions and come up with their own lists, they are more thoughtful about the ways they interact in groups. We keep the posters on the walls, and sometimes I remind students and groups of the norms we agreed on.

I also start classes by explaining to students what is important to me. I say that I do not value speed or people racing through math; I value people showing how they think about the math, and I like creative representations of ideas. I also tell students how important it is to listen to each

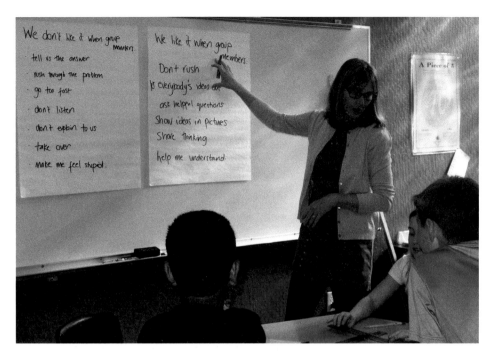

We don't like it when group members:
- tell us the answer
- rush through the problem
- go too fast
- don't listen
- don't explain to us
- take over
- make me feel stupid

We like it when group members:
- Don't rush
- Is everybody's ideas out
- ask helpful questions
- Show ideas in pictures
- Share thinking
- help me understand

A Piece of π

FIGURE 7.11   A teacher generates posters of students' preferred group work behaviors

other's thinking and be respectful of each other. In Chapter Nine I will share a wonderful and specific activity that helps students learn to work well together in groups.

Railside teachers also encouraged group responsibility by approaching a group and asking one student a mathematics question related to the group's work. This follow-up question was always conceptual and could be answered only by the student asked. A student was always chosen at random, and the other students could not help that student. If the student could not answer the question, the teacher would leave the group, telling the other students that they needed to make sure everyone understood, before returning to ask the same student the question again. In the intervening time, it was the group's responsibility to help the student learn the mathematics he or she needed to answer the question. In the following interview two girls, Gita and Brianna, directly link their views of mathematics and their learned responsibility to that particular teacher move:

Interviewer: Is learning math an individual or a social thing?
G: It's like both, because if you get it, then you have to explain it to everyone else. And then sometimes you just might have a group problem and we all have to get it. So I guess both.
B: I think both—because individually you have to know the stuff yourself so that you can help others in your group work and stuff like that. You have to know it so you can explain it to them. Because you never know which one of the four people she's going to pick. And it depends on that one person that she picks to get the right answer. (Railside, year 2)

The students in the preceding extract made the explicit link between teachers asking any group member to answer a question, and being responsible for their group members. They also communicate an interesting social orientation toward mathematics, saying that the purpose in knowing individually is not to be better than others but so "you can help others in your group."

Yet another way the Railside teachers encouraged group responsibility is a method that is shocking to some, but that really communicates the idea that group members are responsible for each other. Occasionally the teachers gave what they called "group tests." Students would take the test individually, but the teachers would take in only one test paper per group (randomly chosen) and grade it. That grade would then be the grade for all the students in the group. This was a very clear communication to students that they needed to make sure all of their group members understood the mathematics.

The students came to Railside School with eight years' experience working individually in math and viewing math as an individual, competitive activity. They learned about a different mathematics and different learning goals at Railside, and they quickly and readily adapted to them. Soon after arriving at the school they started to see math as a collaborative, shared pursuit, all about helping each other and working together. In the first months of our study the high achievers complained to us, saying that they were always having to explain work, but after a few months even their attitude shifted. They started to appreciate being in groups and having the chance to explain their thinking, because they realized it helped their own understanding so much.

Imelda, one of the girls taking calculus in the later years of our study, described the ways in which the social responsibility she learned helped her:

I: I think people look at it as a responsibility, I think it's something they've grown to do, like since we've taken so many math classes. So maybe in ninth grade it's like "Oh my God I don't feel like helping them, I just wanna get my work done, why do we have to take a group test?" But once you get to AP Calc, you're like "Ooh I need a group test before I take a test." So like the more math you take and the more you learn, you grow to appreciate, like "Oh, thank God I'm in a group!" (Imelda, Railside, year 4)

Although it was not the goal of the Railside teachers, in our statistical analyses we found that the students most advantaged by the heterogeneous grouping and the CI approach were the initial high achievers. Their learning accelerated more than other students' at Railside, and they performed at significantly higher levels than the students who went into top tracks at the other schools—in part because the students were explaining work, which took their understanding to new levels, and they were working more multidimensionally. Many of them had come in as fast, procedural workers, and the push to work with more breadth and depth helped their achievement enormously.

The students also developed broader perceptions of the value of different students, and they began to realize that all students could offer something in the solving of problems. As the approach they experienced became more multidimensional, they came to regard each other in more multidimensional ways, valuing the different ways of seeing and understanding mathematics that different students brought to problems. Two of the girls reflected in interviews:

Interviewer: What do you guys think it takes to be successful in math?
A: Being able to work with other people.
E: Be open-minded, listen to everybody's ideas.
A: You have to hear other people's opinions 'cause you might be wrong.
E: You might be wrong 'cause there's lots of different ways to work everything out.
A: 'Cause everyone has a different way of doing things, you can always find different ways to work something out, to find something out.
E: Someone always comes up with a way to do it, we're always like "Oh my gosh, I can't believe you would think of something like that." (Ayana and Estelle, Railside, year 4)

In interviews, the students also told us that they learned to value students from different cultures, classes, and genders because of the mathematics approach used in the school:

R: I love this school, you know? There are schools that are within a mile of us that are completely different—they're broken up into their race cliques and things like that. And at this school everyone's accepted as a person, and they're not looked at by the color of their skin.
Interviewer: Does the math approach help that, or is it a whole school influence?
J: The groups in math help to bring kids together.
R: Yeah. When you switch groups, that helps you to mingle with more people than if you're just sitting in a set seating chart where you're only exposed to the people that are sitting around you, and you don't know the people on the other side of the room. In math you have to talk, you have to voice if you don't know, or voice what you're learning. (Robert and Jon, Railside, year 4)

The mathematics teachers at Railside valued equity very highly, but they did not use special curriculum materials designed to raise issues of gender, culture, or class, as some have recommended (Gutstein, Lipman, Hernandez, & de los Reyes, 1997); instead, they taught students to appreciate the different ways that everyone saw mathematics, and as the classrooms became more multidimensional students learned to appreciate the insights of a wider group of students from different cultures and circumstances.

Many parents worry about high achievers in heterogeneous classes, thinking that low achievers will bring down their achievement, but this does not usually happen. High achievers are often high achievers in the U.S. system because they are procedurally fast. Often these students have not learned to think deeply about ideas, explain their work, or see mathematics from different perspectives because they have never been asked to do so. When they work in groups with different thinkers they are helped, both by going deeper and by having the opportunity to explain work, which deepens their understanding. Rather than groups being lowered by the presence of low achievers, group conversations rise to the level of the highest-thinking students. Neither the high nor the low achievers would be as helped if they were grouped only with similar achieving students.

The students at Railside realized that students were different in what they knew, but they came to value the different strengths of different students, as Zac expressed to us in an interview:

> Everybody in there is at a different level, but what makes the class good is that everybody's at different levels, so everybody's constantly teaching each other and helping each other out.

Two practices I have came to regard as particularly important in promoting equity, and that were central to the responsibility students showed for each other, are justification and reasoning. At Railside, students were required to justify their answers, giving reasons for their methods, at all times. There are many good reasons for this—justification and reasoning are intrinsically mathematical practices (Boaler, 2013c)—but these practices also serve an interesting and particular role in the promotion of equity.

The following interview extract comes from Juan, who at the time was one of the lower achievers in the class. He described the ways he was supported by the practices of justification and reasoning:

> Most of them, they just know what to do and everything. First you're like "Why you put this?" and then if I do my work and compare it to theirs, theirs is like, super different, 'cause they know what to do. I will be like, "Let me copy"; I will be like, "Why you did this?" And then I'd be like, "I don't get it why you got that." And then sometimes the answer's just … "Yeah, he's right and you're wrong." "But like—why?" (Juan, year 2)

Juan made it clear that he was helped by the practice of justification and that he felt comfortable pushing other students to go beyond answers and explain *why* their answers were given, in other words to reason. At Railside, the teachers carefully prioritized the message that every student had two important responsibilities, as the following words, that were on posters around the room, highlight:

> Always give help when needed, always ask for help when you need it

Both of these responsibilities were important in the pursuit of equity, and justification and reasoning emerged as helpful practices in the learning of a wide range of students.

It would be hard to spend years in the classrooms at Railside without noticing that the students were learning to treat each other in more respectful ways than is typically seen in schools, and that ethnic cliques like the ones Robert and Jon described were less evident in the mathematics classrooms than they are in most classrooms. As students worked in math, they were taught to appreciate the contributions of different students, from many different cultural groups and with many different characteristics and perspectives. It seemed to me that the students learned something extremely important through this process—something that would serve them and others well in their future interactions in society. I have termed this *relational* equity (Boaler, 2008), a form of equity that is less about equal scores on tests and more about respect and regard for others, whatever their culture, race, religion, gender, or other characteristic. It is commonly believed that students will learn respect for different people and cultures if they have discussions about such issues or read diverse forms of literature in English or social studies classes. I propose that *all* subjects have something to contribute in the promotion of equity, and that mathematics—often regarded as the most abstract subject, far removed from responsibilities of cultural or social awareness—has an important contribution to make. For the respectful relationships that Railside students developed across cultures and genders were made possible only by a mathematics approach that valued different insights, methods, and perspectives in the collective solving of particular problems.

Conclusion

Equitable, growth mindset teaching is harder than more traditional teaching in which teachers lecture and give short closed questions for practice. It involves teaching broad, open, multidimensional mathematics, teaching students to be responsible for each other, and communicating growth mindset messages to students. This is also the most important and rewarding teaching that a mathematics teacher can do: teachers quickly feel fulfillment and energy from seeing engaged and high-achieving students. I am fortunate to have worked with so many teachers who are committed to equity and to growth mindset teaching and grouping—even if they did not use these words—and this chapter has shared some of the insights I have gained after many years of working and conducting research with these highly successful teachers. I have saved my favorite strategy for encouraging good group work for the book's final chapter, Chapter Nine, which will explain all of the norms and methods I recommend for creating a growth mindset math class.

CHAPTER 8

Assessment for a Growth Mindset

The complex ways in which children understand mathematics are fascinating to me. Students ask questions, see ideas, draw representations, connect methods, justify, and reason in all sorts of different ways. But recent years have seen all of these different nuanced complexities of student understanding reduced to single numbers and letters that are used to judge students' worth. Teachers are encouraged to test and grade students, to a ridiculous and damaging degree; and students start to define themselves—and mathematics—in terms of letters and numbers. Such crude representations of understanding not only fail to adequately describe children's knowledge, in many cases they misrepresent it.

In the United States, students are overtested to a degree that is nothing short of remarkable, particularly in mathematics. For many years students have been judged by narrow, procedural mathematics questions presented with multiple-choice answers. The knowledge needed for success on such tests is so far from the adaptable, critical, and analytical thinking needed by students in the modern world that leading employers such as Google have declared they are no longer interested in students' test performance, as it in no way predicts success in the workplace (Bryant, 2013).

One critical principle of good testing is that it should assess what is important. For many decades in the United States, tests have assessed what is easy to test instead of important and valuable mathematics. This has meant that mathematics teachers have had to focus their teaching on narrow procedural mathematics, not the broad, creative, and growth mathematics that is so important. The new common core assessments promise something different, with few multiple-choice questions and more assessments of problem solving, but they are being met with considerable opposition from parents.

The damage does not end with standardized testing, for math teachers are led to believe they should use classroom tests that mimic low-quality standardized tests, even when they know the tests assess narrow mathematics. They do this to help prepare students for later success. Some teachers, particularly at the high school level, test weekly or even more frequently. Mathematics teachers feel the need to test regularly, more than in any other subject, because they have come to believe that mathematics is about performance, and they usually don't consider the negative role that tests play in shaping students' views of mathematics and themselves. Many mathematics teachers I know start the year or a math class with a test, which gives a huge performance message to students on the first day of class—a time when it is so important to be giving growth messages about mathematics and learning.

Finland is one of the highest-scoring countries in the world on international mathematics tests, yet students there do not take any tests in school. Instead, teachers use their rich understanding of their students' knowledge gained through teaching to report to parents and make judgments about work. In a longitudinal study I conducted in England, students worked on open-ended projects for three years (ages 13 through 16), leading to national standardized examinations. They did not take tests in class, nor was their work graded. Students encountered short questions assessing procedures only in the last few weeks before the examination, as the teachers gave them examination papers to work through. Despite the students' lack of familiarity with answering examination questions or working under timed conditions of any kind, they scored at significantly higher levels than a matched cohort of students who had spent three years working through questions similar to the national exam questions and taking frequent tests. The students from the problem-solving school did so well in the standardized national exam because they had been taught to believe in their own capabilities; they had been given helpful, diagnostic information on their learning; and they had learned that they could solve any question, as they were mathematical problem solvers.

As part of my research study, I was given access to the students' completed national examination papers (GSCE), which the examination board kept under lock and key in England. The examination board granted my unusual request, as they agreed it would be helpful for the advancement of research knowledge. I spent a day in a cupboard-sized room, with no windows, deep in the offices of the examination board, recording and analyzing all the students' examination responses. This was very enlightening. I found that the students who had worked through open-ended projects tackled significantly more of the examination questions, trying to work them out whether they recognized the problem or not—an important and valuable practice that all students should learn. They were also more successful on the questions they attempted, even when the questions assessed a standard method they had never been taught. I divided all the questions into the categories of procedural and conceptual and found that students from the two schools scored at the same level on procedural questions, which involved the simple use of a standard method. The project-based students achieved at significantly higher levels on the conceptual questions, which required more thought. The higher achievement in tests among students who do not take tests in school may seem counterintuitive, but the new research on the brain and learning makes sense of this result. Students with no experience of examinations and tests can score at high levels because the most important preparation we can give students is a growth mindset, positive beliefs about their own ability, and problem-solving mathematical tools that they are prepared to use in any mathematical situation.

The testing regime of the last decade has had a large negative impact on students, but it does not end with testing; the communication of grades to students is similarly negative. When students are given a percentage or grade, they can do little else besides compare it to others around them, with half or more deciding that they are not as good as others. This is known as "ego feedback," a form of feedback that has been found to damage learning. Sadly, when students are given frequent test scores and grades, they start to see themselves as those scores and grades. They do not regard the scores as an indicator of their learning or of what they need to do to achieve; they see them as indicators of who they are as people. The fact that U.S. students commonly describe themselves saying "I'm an A student" or "I'm a D student" illustrates how students define themselves by grades. Ray McDermott wrote a compelling paper about the capturing of a child by a learning disability, describing the ways a student who thought and worked differently was given a label and was then defined by that label (McDermott, 1993). I could give a similar argument about the capturing of students by grades and test scores. Students describe themselves as A or D students because they have grown up in a performance culture that has, for a long time, valued frequent testing and grading, rather than persistence, courage, or problem solving. The traditional methods of assessing students that have been used across the United States for decades were designed in a less enlightened age (Kohn, 2011) when it was believed that grades and test scores would motivate students and that the information they provided on students' achievement would be useful. Now we know that grades and test scores demotivate rather than motivate students, and that they communicate fixed and damaging messages to students that result in lower classroom achievement.

In studies of grading and alternatives to grading, researchers have produced consistent results. Study after study shows that grading reduces the achievement of students. Elawar and Corno, for example, contrasted the ways teachers responded to math homework in sixth grade, with half of the students receiving grades and the other half receiving diagnostic comments without a grade (Elawar & Corno, 1985). The students receiving comments learned twice as fast as the control group, the achievement gap between male and female students disappeared, and student attitudes improved.

Ruth Butler also contrasted students who were given grades for classwork with those who were given diagnostic feedback and no grades (Butler, 1987, 1988). Similar to Corno and Elawar, the students who received diagnostic comments achieved at significantly higher levels. What was fascinating in Butler's study was that she then added a third condition, which gave students grades *and* comments—as this could be thought of as the best of both worlds. However, this showed that the students who received grades only and those who received grades and comments scored equally badly, and the group that achieved at significantly higher levels was the comment-only group—when students received a grade and a comment, they cared about and focused on only the grade. Butler found that both high-achieving (the top 25% GPA) and low-achieving (the bottom 25% GPA) fifth and sixth graders suffered deficits in performance and motivation in both graded conditions, compared with the students who received only diagnostic comments.

Pulfrey, Buchs, and Butera (2011) followed up on Butler's study, replicating her finding—showing again that students who received grades as well as students who received grades and comments both underperformed and developed less motivation than students who received only comments. They also found that students needed only to *think* they were working for a grade to lose motivation, resulting in lower levels of achievement.

The move from grades to diagnostic comments is an important one, and is a move that allows teachers to give students an amazing gift—the gift of their knowledge and insights about ways to improve. Teachers, quite rightly, worry about the extra time this can take, as good teachers already work well beyond the hours they are paid for. My recommended solution is to assess less; if teachers replaced grading weekly with diagnostic comments given occasionally, they could spend the same amount of time, eliminate the fixed mindset messages of a grade, and provide students with insights that would propel them onto paths of higher achievement. Later in this chapter I share some accounts from different teachers I have worked with who shifted their modes of assessment without investing extra time. They describe their work and the impact it had on their students.

Race to Nowhere

Race to Nowhere is a documentary film highlighting the stress students are placed under in U.S. schools (see Figure 8.1). It was released a few years ago to widespread attention and acclaim—the

New York Times, for example, described it as a "must see" movie. Soon after the film's release it was playing to packed movie theatres and school halls across the country. The film showed the damaging effects of testing, grading, homework, and overscheduling on students' health and well-being. The *Race to Nowhere* campaign has continued to gather support from tens of thousands of educators and parents. When I watched *Race to Nowhere,* throughout the film I saw math as the main source of students' stress and anxiety. The film features the sad story of a high school girl, Devon Marvin, who had always done well in math. She was a highly motivated young girl who saw math as part of her identity. One day she received an F on a math test and—tragically—she committed suicide. For Devon and for many other students, the grade she received did not communicate a message about an area of math she needed to work on in her growth learning path; instead, it gave her a message about who she was as a person—she was now an F student. This idea was so crushing to her, she decided to take her own life.

FIGURE 8.1 *Race to Nowhere* poster
Source: Image courtesy of Reel Link Films.

When we give assessments to students, we create an important opportunity. Well-crafted tasks and questions accompanied by clear feedback offer students a growth mindset pathway that helps them to know that they can learn to high levels and, critically, how they can get there. Unfortunately, most systems of assessment in U.S. classrooms do the opposite, communicating information to students that causes many of them to think they are a failure and they can never learn math. I have worked with teachers in recent years who have shifted their methods of assessment from standard tests with grades and scores to assessments focused on giving students the information they need in order to learn well, accompanied with growth mindset messages. This resulted in dramatic changes in their classroom environments. Math anxiety, formerly commonplace among students, disappeared and was replaced by student self-confidence, which led to higher levels of motivation, engagement, and achievement. I will share in this chapter some of the shifts we need to make in classrooms to replace fixed mindset testing with growth mindset assessment that empowers learners.

The director of *Race to Nowhere,* Vicky Abeles, has released a sequel entitled *Beyond Measure.* In her work developing the sequel and in her interviews with students and parents across the United States she has realized that math is the subject most in need of change; the subject that, more than any other, ends students' dreams of college and even of high school graduation. This prompted Vicky to dedicate an entire new film to the issues in math. In the new film, an extract of which can be seen online (https://www.youcubed.org/the-american-math-crisis-forthcoming-documentary/), she captures the work I have been doing over the last few years with teachers in one school district where there was widespread math failure (for more information, visit http://www.racetonowhere.com/american-math-crisis). In Vista School District in San Diego, as in many other urban districts in the United States, over half of students were failing algebra and then entering a cycle of repeated failure. But math failure does not end with math. The numbers of students finishing the high schools in Vista with what is known as "A-G completion" (of required high school courses), or college readiness, was a shocking 24%. Fortunately Vista had an innovative superintendent and director of mathematics—Devin Vodicka and Cathy Williams, respectively. They knew that changes needed to be made, and they were prepared to invest time and energy in making them happen. I spent the next year working with all the middle school teachers in the district, conducting professional development on ways to teach mathematics well, to group students for success and to assess for a growth mindset.

The district math leader, Cathy Williams, had decided that all middle school teachers across the district would work with me over the year—whether they wanted to or not. This meant that when I first met the group of teachers they had very mixed levels of motivation to change. I still remember Frank, an older man, close to retirement, who was not about to change the traditional teaching model he had used for his whole career. He sat unenthusiastically through the first few sessions with me. But gradually Frank started to pick up the excitement of the other teachers and recognize the importance of the research I shared, and he realized he could give his students a changed and better mathematics future. I still remember vividly the session late in the year when he rushed excitedly into the room, telling us all that he had spent his weekend, with his wife, making a life-sized graph on a tarp and how wonderful his math lesson had been when he asked the students to walk around the graph, helping them to understand the meaning of graphical relations. I saw the same excitement shift in all of the teachers, as they tried new ideas with their students and saw their increased engagement.

I am a strong supporter of teachers, and I know that the No Child Left Behind era stripped the professionalism and enthusiasm of many teachers as they were forced (and I choose that word carefully) to use teaching methods that they knew to be unhelpful. An important part of my work with teachers now is to help them regain their sense of professionalism. In the professional development sessions I conducted in Vista, the teachers started to see themselves as creators again, people who could design teaching environments infused with their own ideas for creative, engaging math. This is a much more rewarding role for teachers and one that I encourage in all teachers I work with. During this process I watched the teachers come alive. The energy in the room increased every day we spent together. Over the course of a year the teachers shifted their teaching from worksheet math to inquiry-based math; they chose to eliminate tracks so that they could teach all students that they could be high achievers, and they changed their methods of assessment from fixed to growth assessment. I have seen this shift happen with many teachers with whom I have worked. It comes about when teachers are treated as the professionals they are and are invited to use their own judgment, helped by research ideas, to create positive learning and assessment experiences for their students.

FIGURE 8.2   Delia, *Beyond Measure*
Source: Image courtesy of Reel Link Films.

In her newer film, Vicki Abeles and her team interviewed some of the middle school students in the district, hearing from them about the changes that were happening in their classrooms after the professional development year. One girl, Delia, talked about getting an F for her homework in the previous year, and how it had caused her to stop trying in math and—shockingly—all of her classes across the school (see Figure 8.2). In the interview she poignantly says, "When I saw the F on my paper I felt like a nothing. I was failing in that class so I thought I may as well fail in all my other classes too. I didn't even try." Later in the film she talks about the change in her math class and how she now feels encouraged to do well. "I hated math," she says. "I absolutely hated it—but now I have a connection with math, I'm open, I feel like I'm alive, I'm more energetic."

Delia's use of the word "open" in describing how she felt about math echoes a sentiment I hear frequently from students when they are taught mathematics without the impending fear of low test scores and grades. But it goes further than assessment—when we teach creative, inquiry math, students feel an intellectual freedom that is powerful. In interviews with third graders who experienced number talks in class (Chapter Four), I asked the students how they feel about number talks. The first thing young Dylan said in the interview was, "I feel free." He went on to describe how the valuing of different mathematics strategies allowed him to feel he could work with mathematics in any way he wanted, to explore ideas and learn about numbers. The students' use of words such as "free" and "open" demonstrate the difference that is made when students work on growth mindset mathematics; this goes well beyond math achievement to an intellectual empowerment that will affect students throughout their lives (Boaler, 2015a).

The perceptions students develop about their own potential affect their learning, their achievement, and, of equal importance, their motivation and effort—as Delia describes in the film. When she got an F in math, she gave up not only in math but also in all of her other classes, as she felt like a failure. This is not an unusual response to grading. When students are given scores that tell them they rank below other students, they often give up on school, deciding that they will never be able to learn, and they take on the identity of an underperforming student. The grades and scores given to students who are high achieving are just as damaging. Students develop the idea that they are an "A student" and start on a precarious fixed mindset learning path that makes them avoid harder work or challenges for fear that they will lose their A label. Such students often are devastated if they get a B or lower, for any of their work.

After a recent presentation to teachers on the negative impact of grading, an experienced high school teacher rushed up to talk to me. He said he had been a high school math teacher for over 20 years, always grading students, until the last year, when he had given up grading. He said the impact of the change was phenomenal: the whole classroom turned into an open learning space, with students working harder and achieving at higher levels. Instead of grading, he said, he gave students assessments where they answered as many problems as they could. When they reached a point when the questions became difficult and they felt they couldn't answer them, he asked them to draw a line across the paper and answer the rest of the questions with the help of a book. When students finished the assessment, the work they had done beneath the line became the work they all discussed in class. The teacher said that the assessments, which communicated wonderful growth messages to students, also gave him the best information he had ever had on the mathematics he was teaching, with a quick and easy way to see what students were struggling with and what should be the topic of class discussions.

In another research study on grading, Deevers found that students who were not given scores but instead given positive constructive feedback were more successful in their future work (Deevers, 2006). He also, sadly, found that as students got older teachers gave less constructive feedback and more fixed grading. He found a clear and unsurprising relationship between teachers' assessment practices and student beliefs, as students' beliefs about their own potential and the possibility of improving their learning declined steadily from fifth to twelfth grade (Deevers, 2006). The traditional high school assessment culture of math teachers believing that they need to give frequent fixed mindset tests and grades to students has prevailed for decades, so I was particularly pleased to hear from that experienced high school teacher who had changed his approach, created "open" classrooms, and immediately saw changes in students' motivation and learning.

We want students to be excited about and interested in their learning. When students develop interest in the ideas they are learning, their motivation and their achievements increase. There is a large body of research that has studied two types of motivation. Intrinsic motivation comes from interest in the subject and ideas you are learning; extrinsic motivation is the motivation provided by the thought of getting better scores and grades. Because mathematics has been taught for decades as a performance subject, the students who are most motivated in math classrooms are usually those who are extrinsically motivated. One result is that usually the only students who feel positive about math class are the ones getting high scores and grades. Most of the teachers who believe in grades use them because they think they motivate students to achieve. They do motivate some students—those who would probably achieve at high levels anyway—but they de-motivate the rest. Unfortunately, the extrinsic motivation that the high-achieving students develop is not helpful in the long term. Study after study shows that students who develop intrinsic motivation achieve at higher levels than those who develop extrinsic motivation (Pulfrey, Buchs, & Butera, 2011; Lemos & Veríssimo, 2014), and that intrinsic motivation spurs students to pursue subjects to higher levels and to stay in subjects rather than drop out (Stipek, 1993).

I saw the difference in the effects of intrinsic and extrinsic motivation in my own daughter when she entered fifth grade. She had been attending our local public elementary school, which used no grading and only minimal testing, so up until fifth grade she had only ever received feedback on her actual work, and she had a wonderful form of intrinsic motivation that I watched develop, as she would come home and excitedly tell me about the ideas she was learning. In fifth grade she had a highly accomplished teacher who filled the classroom with rich and engaging activities, but he graded all of the students' work. He did so, he told me, because the local middle schools graded everything, and he wanted to prepare his students for that experience. Alfie Kohn describes this kind of approach as BGUTI—"better get used to it." Damaging practices are used in schools because teachers know that students will experience them later and want them to get used to them. During that fifth-grade year I saw a massive change in my daughter, who suddenly started to care—and worry—only about grades. She turned her attention away from the ideas she was learning and worried constantly about the grade she might get for her work. Alfie Kohn (2011) quotes from a student named Claire who described a similar shift:

> I remember the first time that a grading rubric was attached to a piece of my writing.... Suddenly all the joy was taken away. I was writing for a grade—I was no longer exploring for me. I want to get that back. Will I ever get that back?

As Claire describes it, her sense of exploration and joy disappeared. For my own daughter there was a happy ending, as in sixth grade she moved to a school that did not grade, and I saw her interest in learning return. But for many students this does not happen, and they move through middle schools getting more and more grades and feeling less and less motivated by the ideas they are learning.

I give more detail in *What's Math Got to Do with It?* on the damage caused by grading and testing, for all achievement levels, and the research evidence pointing to its negative impact (Boaler, 2015a). I also recommend Alfie Kohn's highly readable articles and books on the impact of traditional assessment practices (Kohn, 1999, 2000). In the remainder of this chapter I will focus on the ways we can assess students with a growth orientation, giving growth information and messaging to students that sets them on an informed and positive pathway toward success. This is one of the most important changes a teacher can make in her classrooms.

Assessment for Learning

A few years ago, two professors from England—Paul Black and Dylan Wiliam—conducted a meta-analysis of hundreds of research studies on assessment. They found something amazing: a form of assessment so powerful that if teachers shifted their practices and used it, it would raise the achievement of a country, as measured in international studies, from the middle of the pack to a place in the top five. (Sir Paul Black and Professor Dylan Wiliam were both good colleagues of mine at London University; Paul Black was also my dissertation advisor and mentor.) Black and Wiliam found that if teachers were to use what is now called "assessment for learning," the positive impact would be far greater than that of other educational initiatives such as reductions in class size (Black, Harrison, Lee, Marshall, & Wiliam, 2002; Black & Wiliam, 1998a, 1998b). They published their findings in a small booklet that sold over 20,000 copies in the first few weeks in England. Assessment for learning is now a national initiative in many countries; it has a huge research evidence base, and it communicates growth mindset messages to students.

A little background will be helpful. There are two types of assessment—formative and summative. Formative assessment informs learning and is the essence of assessment for learning or A4L. Formative assessments are used to find out where students are in their learning so that teachers and students can determine what they need to know next. The purpose of summative assessment, in contrast, is to summarize a student's learning—to give a final account of how far a student has gotten, as an end point. One problem in the United States is that many teachers use summative assessment formatively; that is, they give students an end score or grade when they are still learning the materials. In mathematics classrooms teachers often use summative tests weekly and then move on to the next subject without waiting to see what the tests reveal. In A4L, students become knowledgeable about what they know, what they need to know, and ways to close the gap between the two. Students are given information about their flexible and growing learning pathways that contributes to their development of a growth mathematics mindset.

In the weeks and months that students are learning in a course, it is very important to assess formatively, not summatively. Further, the A4L approach, which can also be thought of as assessing for a growth mindset, offers a range of strategies and methods.

One important principle of A4L is that it teaches students responsibility for their own learning. At its core A4L is about empowering students to become autonomous learners who can self-regulate and determine what they most need to learn and who know ways to improve their learning. Assessment for learning can be thought of as having three parts: (1) clearly communicating to students what they have learned, (2) helping students become aware of where they are in their learning journey and where they need to reach, and (3) giving students information on ways to close the gap between where they are now and where they need to be (see Figure 8.3).

The approach is called assessment *for* learning rather than assessment *of* learning because the information teachers and students get from A4L helps teachers make their instruction more effective and helps students learn, to the greatest extent possible. Teachers who use A4L spend less time telling students their achievement and more time empowering students to take control

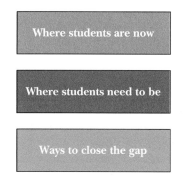

FIGURE 8.3   Assessment for Learning

of their learning pathways. A teacher in England who shifted to A4L practices reflected that it "made me focus less on myself but more on the children" (Black et al., 2002). He developed more confidence as a teacher because of powerful strategies to empower his students to take their own learning forward.

Developing Student Self-Awareness and Responsibility

The most powerful learners are those who are reflective, who engage in metacognition—thinking about what they know—and who take control of their own learning (White & Frederiksen, 1998). A major failing of traditional mathematics classes is that students rarely have much idea of what they are learning or where they are in the broader learning landscape. They focus on methods to remember but often do not even know what area of mathematics they are working on. I have visited math classes many times and stopped at students' desks to ask them what they are working on. Often students simply tell me the question they are working on. Many of my interactions have gone something like this:

JB: What are you working on?
Student: Exercise 2.
JB: So what are you actually doing? What math are you working on?
Student: Oh, I'm sorry—question 4.

Students are often not thinking about the area of mathematics they are learning, they do not have an idea of the mathematical goals for their learning, and they expect to be passively led through work, with teachers telling them whether they are "getting it" or not. Alice White, an assessment expert, likens this situation to workers on a ship who are given jobs to do each day but don't have any idea where the ship is travelling to.

One research study, conducted by Barbara White and John Frederiksen (White & Frederiksen, 1998), powerfully illustrated the importance of reflection. The researchers studied twelve classes of seventh-grade students learning physics. They divided the students into experimental and control groups. All groups were taught a unit on force and motion. The control groups then spent some of each lesson discussing the work, whereas the experimental group spent some of each lesson engaging in self- and peer assessment, considering criteria for the science they were learning. The results of the study were dramatic. The experimental groups outperformed the control groups on three different assessments. The previously low-achieving students made the greatest gains. After they spent time considering the science criteria and assessing themselves against them, they began to achieve at the same levels as the highest achievers. The middle school students even scored at higher levels than AP physics students on tests of high school physics. The researchers concluded that a large part of the students' previous low achievement came not from their purported lack of ability but from the fact that previously they had not known what they should really be focusing upon.

This is sadly true for many students. It is so important to communicate to students what they should be learning. This both helps the students know what success is and starts a self-reflection process that is an invaluable tool for learning.

There are many strategies for encouraging students to become more aware of the mathematics they are learning and their place in the learning process. Beginning here, I will share nine of my favorites.

1. Self-Assessment

The two main strategies for helping students become aware of the math they are learning and their broader learning pathways are self- and peer assessment. In self-assessment, students are given clear statements of the math they are learning, which they use to think about what they have learned and what they still need to work on. The statements should communicate mathematics content such as "I understand the difference between mean and median and when each should be used" as well as mathematical practices such as "I have learned to persist with problems and keep going even when they are difficult." If students start each unit of work with clear statements about the mathematics they are going to learn, they start to focus on the bigger landscape of their learning journeys—they learn what is important, as well as what they need to work on to improve. Studies have found that when students are asked to rate their understanding of their work through self-assessment, they are incredibly accurate at assessing their own understanding, and they do not over- or underestimate it (Black et al., 2002).

Self-assessment can be developed at different degrees of granularity. Teachers could give students the mathematics in a lesson or show them the mathematics from a longer period of time, such as a unit or even a whole term or semester. Examples of self-assessment criteria for shorter and longer time periods are provided here. In addition to receiving the criteria, students need to be given time to reflect upon their learning, which they can do during a lesson, at the end of a lesson, or even at home.

The self-assessment example in Exhibit 8.1 comes from a wonderful third-grade teacher I have worked with, Lori Mallet. Lori attended a summer professional development workshop I taught where we considered all the ways of encouraging a growth mindset. In her self-assessment example she offers three options for the students to choose from.

Rather than giving words for students to reflect upon, some teachers, particularly of younger children, use smiley faces such as those in Figure 8.4.

Both options prompt students to consider what they have learned and what they need to learn.

My second example of self-assessment comes from Lisa Henry, an expert high school teacher from Brookland, Ohio. Lisa has been teaching high school mathematics for 23 years. Four years ago Lisa became dissatisfied with grading. She knew that her grades did not represent what the students knew. Lisa moved to assessing students against criteria that she shared with the students. Lisa kindly shares with others the self-assessment statements she wrote for the whole of her algebra 1 course (see Exhibit 8.2). Now that students are assessing themselves against criteria and Lisa is assessing students by deciding what they know and don't know, instead of an overall grade, she says she knows a lot more about students' knowledge and understanding.

Self-Assessment: Polygons

	I can do this independently and explain my solution path(s) to my classmate or teacher.	I can do this independently.	I need more time. I need to see an example to help me.
Draw lines and line segments with given measurements.			
Draw parallel lines and line segments.			
Draw intersecting lines and line segments.			
Create a polygon with a given perimeter.			
Create a square or rectangle with a given area.			
Create an irregular shape whose area can be solved by cutting it into rectangles or squares.			

Source: From Lori Mallet

E x h i b i t   8 . 1

| I've got it! | I'm struggling and learning | I need some help |

F I G U R E   8 . 4   Self-reflection faces

Algebra 1 Self-Assessment

Unit 1 – Linear Equations and Inequalities

☐ I can solve a linear equation in one variable.

☐ I can solve a linear inequality in one variable.

☐ I can solve formulas for a specified variable.

☐ I can solve an absolute value equation in one variable.

☐ I can solve and graph a compound inequality in one variable.

☐ I can solve an absolute value inequality in one variable.

Unit 2 – Representing Relationships Mathematically

☐ I can use and interpret units when solving formulas.

☐ I can perform unit conversions.

☐ I can identify parts of an expression.

☐ I can write the equation or inequality in one variable that best models the problem.

☐ I can write the equation in two variables that best model the problem.

☐ I can state the appropriate values that could be substituted into an equation and defend my choice.

☐ I can interpret solutions in the context of the situation modeled and decide if they are reasonable.

☐ I can graph equations on coordinate axes with appropriate labels and scales.

☐ I can verify that any point on a graph will result in a true equation when their coordinates are substituted into the equation.

☐ I can compare properties of two functions graphically, in table form, and algebraically.

Unit 3 – Understanding Functions

☐ I can determine if a graph, table, or set of ordered pairs represents a function.

☐ I can decode function notation and explain how the output of a function is matched to its input.

☐ I can convert a list of numbers (a sequence) into a function by making the whole numbers the inputs and the elements of the sequence the outputs.

Exhibit 8.2

☐ I can identify key features of a graph, such as the intercepts, whether the function is increasing or decreasing, maximum and minimum values, and end behavior, using a graph, a table, or an equation.

☐ I can explain how the domain and range of a function is represented in its graph.

Unit 4 – Linear Functions

☐ I can calculate and interpret the average rate of change of a function.

☐ I can graph a linear function and identify its intercepts.

☐ I can graph a linear inequality on a coordinate plane.

☐ I can demonstrate that a linear function has a constant rate of change.

☐ I can identify situations that display equal rates of change over equal intervals and can be modeled with linear functions.

☐ I can construct linear functions from an arithmetic sequence, graph, table of values, or description of the relationship.

☐ I can explain the meaning (using appropriate units) of the slope of a line, the y-intercept, and other points on the line when the line models a real-world relationship.

Unit 5 – Systems of Linear Equations and Inequalities

☐ I can solve a system of linear equations by graphing.

☐ I can solve a system of linear equations by substitution.

☐ I can solve a system of linear equations by the elimination method.

☐ I can solve a system of linear inequalities by graphing.

☐ I can write and graph a set of constraints for a linear-programming problem and find the maximum and/or minimum values.

Unit 6 – Statistical Models

☐ I can describe the center of the data distribution (mean or median).

☐ I can describe the spread of the data distribution (interquartile range or standard deviation).

☐ I can represent data with plots on the real number line (dot plots, histograms, and box plots).

E x h i b i t   8 . 2   (Continued)

☐ I can compare the distribution of two or more data sets by examining their shapes, centers, and spreads when drawn on the same scale.

☐ I can interpret the differences in the shape, center, and spread of a data set in the context of a problem, and can account for effects of extreme data points.

☐ I can read and interpret the data displayed in a two-way frequency table.

☐ I can interpret and explain the meaning of relative frequencies in the context of a problem.

☐ I can construct a scatter plot, sketch a line of best fit, and write the equation of that line.

☐ I can use the function of best fit to make predictions.

☐ I can analyze the residual plot to determine whether the function is an appropriate fit.

☐ I can calculate, using technology, and interpret a correlation coefficient.

☐ I can recognize that correlation does not imply causation and that causation is not illustrated on a scatter plot.

Unit 7 – Polynomial Expressions and Functions

☐ I can add and subtract polynomials.

☐ I can multiply polynomials.

☐ I can rewrite an expression using factoring.

☐ I can solve quadratic equations by factoring.

☐ I can sketch a rough graph using the zeroes of a quadratic function and other easily identifiable points.

Unit 8 – Quadratic Functions

☐ I can use completing the square to rewrite a quadratic expression into vertex form.

☐ I can graph a quadratic function, identifying key features such as the intercepts, maximum and/or minimum values, symmetry, and end behavior of the graph.

☐ I can identify the effect of transformations on the graph of a function with and without technology.

☐ I can construct a scatter plot, use technology to find a quadratic function of best fit, and use that function to make predictions.

E x h i b i t   8 . 2   (Continued)

Unit 9 – Quadratic Equations

- ☐ I can explain why sums and products are either rational or irrational.
- ☐ I can solve quadratic equations by completing the square.
- ☐ I can solve quadratic equations by finding square roots.
- ☐ I can solve quadratic equations by using the quadratic formula.

Unit 10 – Relationships That Are Not Linear

- ☐ I can apply the properties of exponents to simplify algebraic expressions with rational exponents.
- ☐ I can graph a square root or cube root function, identifying key features such as the intercepts, maximum and/or minimum values, and end behavior of the graph.
- ☐ I can graph a piecewise function, including step and absolute value functions, identifying key features such as the intercepts, maximum and/or minimum values, and end behavior of the graph.

Unit 11 – Exponential Functions and Equations

- ☐ I can demonstrate that an exponential function has a constant multiplier over equal intervals.
- ☐ I can identify situations that display equal ratios of change over equal intervals and can be modeled with exponential functions.
- ☐ I can use graphs or tables to compare the rates of change of linear, quadratic, and exponential functions.
- ☐ I can rewrite exponential functions using the properties of exponents.
- ☐ I can interpret the parameters of an exponential function in real-life problems.
- ☐ I can graph exponential functions, identifying key features such as the intercepts, maximum and/or minimum values, asymptotes, and end behavior of the graph.
- ☐ I can construct a scatter plot, use technology to find an exponential function of best fit, and use that function to make predictions.

Source: By Lisa Henry

Exhibit 8.2   (Continued)

2. Peer Assessment

Peer assessment is a similar strategy to self-assessment, as it also involves giving students clear criteria for assessment, but they use it to assess each other's work rather than their own. When students assess each other's work they gain additional opportunities to become aware of the mathematics they are learning and need to learn. Peer assessment has been shown to be highly effective, in part because students are often much more open to hearing criticism or a suggestion for change from another student, and peers usually communicate in ways that are easily understood by each other.

One of my favorite methods of peer assessment is the identification of "two stars and a wish" (see Exhibit 8.3). Students are asked to look at their peers' work and, with or without criteria, to select two things done well and one area to improve on.

When students are given information that communicates clearly what they are learning, and they are asked, at frequent intervals, to reflect on their learning, they develop responsibility for

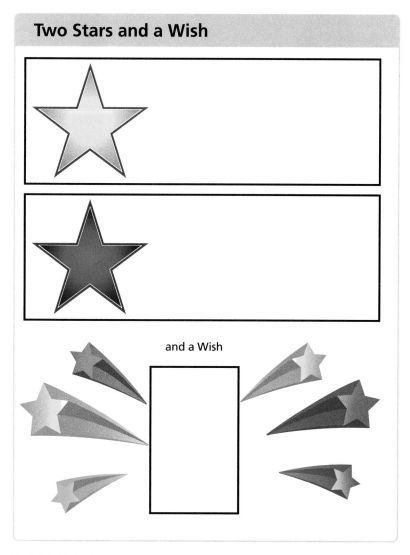

Exhibit 8.3

their own learning. Some people refer to this as inviting students into the guild—giving students the powerful knowledge—knowledge that usually only teachers hold—which allows them to take charge of their learning and be successful.

3. Reflection Time

An effective way for students to become knowledgeable about the ideas they are learning is to provide some class time for reflection. Ask students at the end of a lesson to reflect using questions such as those in Exhibit 8.4.

Exhibit 8.4

4. Traffic Lighting

This is a class activity that prompts reflection for students and gives important information to teachers. There are many different varieties of the traffic lighting activity, but they all involve students using the colors red, yellow, and green (see Figure 8.5) to indicate whether students understand something, partially understand something, or need more work on something. Some teachers distribute different colored paper cups that students put on their desks as they are learning. Students who need a teacher to stop and review put the red cup on their desks, those who feel the lesson is going too fast show the yellow cup, or teachers may come up with variations on this theme. At first some teachers found that students were reluctant to show cups, but after they found how useful it was to them, they used them readily. Some teachers ask tables with a green cup displayed to explain an idea to the rest of the class. This helps students and teachers enormously, as teachers get feedback on their teaching in real time instead of at the end of a unit or piece of work, when it is too late to do anything about it. Instead of cups, teachers can use laminated pieces of colored paper and hole punch them so that they can be placed onto a ring.

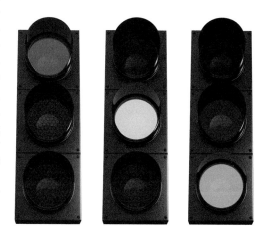

FIGURE 8.5 Traffic lights

5. Jigsaw Groups

In jigsaw groups, students working together become experts on a particular phenomenon, new method, or an interesting reading. Then the groups divide up and new groups are formed so that every group has a member with different expertise. The different group members then teach each other the new knowledge they have learned, working as an expert. You need at least four areas of expertise for this to work, so that when group members move to other groups they all have something different to teach each other about. A class of 32 students with eight groups could work as shown in Figure 8.6.

In Chapter Six I suggested a jigsaw activity in which students became experts on inspirational people who serve to banish stereotypical views on who can achieve in mathematics.

Another example of a jigsaw activity encourages students to understand the connections in algebra between graphs, tables of values, terms of an equation, and patterns. In this example, teachers give out four patterns, such as those in Exhibits 8.5 through 8.8, and ask groups of students to make a poster illustrating how they see the shapes growing, showing a table of values, a graph of the equation, and the pattern generalized and modeled by an equation. Each group member becomes an expert on the multiple representations of their group's pattern. The teacher

Classroom Jigsaw Expert   Classroom Jigsaw Share

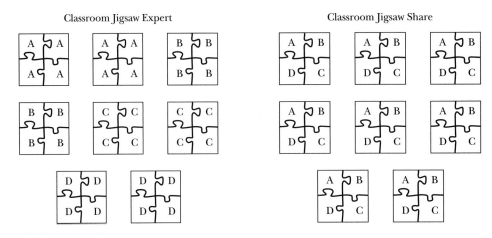

FIGURE 8.6 Jigsaw groups

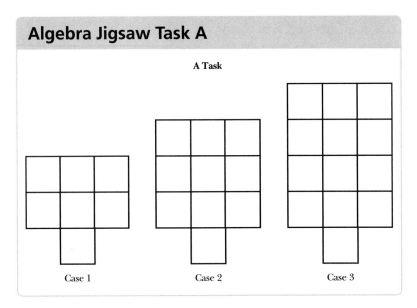

Algebra Jigsaw Task A

A Task

Case 1   Case 2   Case 3

Exhibit 8.5

then asks one member of each group to move to a table with a group member from each of the other tasks. Each group member then shares their expertise about their task with the other group members. The groups then discuss the similarities and differences in their algebraic patterns and representations.

When students become experts and have the responsibility of teaching other members of the class, they are again encouraged to take responsibility for new knowledge they are learning.

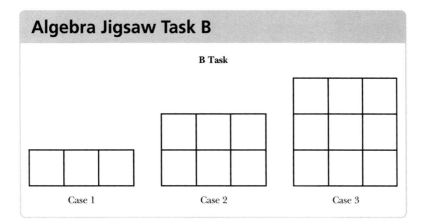

Exhibit 8.6

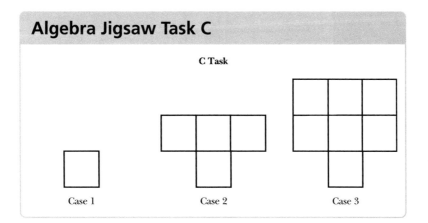

Exhibit 8.7

6. Exit Ticket

An exit ticket is a piece of paper that you give to students at the end of class that asks them to talk about their learning in that class (see Exhibit 8.9). Before they leave the room, they complete the ticket and give it to you. This is another time for students to reflect, helping their learning and giving the teacher really valuable information on students' learning and ideas for the next lesson.

7. Online Forms

An effective strategy I have seen teachers use is to ask students to complete an online form, in real time during the lesson, sending ideas to the teacher's computer. Students can be asked to

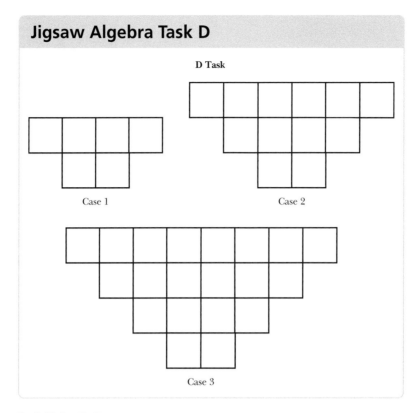

Jigsaw Algebra Task D

D Task

Case 1   Case 2

Case 3

Exhibit 8.8

give comments or thoughts on the lesson. Students who do not normally participate verbally will be more willing to share their thoughts online. There are many ways these can be used, such as asking students to send reflections, having students vote on something, or as a way to give the teacher red, yellow, or green indicators that are not visible to other students.

8. Doodling

As I noted in Chapter Four, brain science tells us that the most powerful learning occurs when we use different pathways in the brain. The classroom implications of this finding are huge—they go beyond assessment practices. They tell us that mathematics learning, particularly the formal abstract mathematics that takes up a lot of the school curriculum, is enhanced when students are using visual and intuitive mathematical thinking, connected with numerical thinking. A really good way to encourage this is through drawing ideas (Figure 8.7).

Instead of asking students to write down only what they understand, as a reflection or after the lesson, ask them to show their understanding in a sketch, cartoon, or doodle. If you would

Exit Ticket

Exit Ticket		Name
		Date
Three things I learned today …	Two things I found interesting …	One question I have …

Exit Ticket

Exhibit 8.9

like to see, and perhaps show your students, some highly impressive and entertaining doodles of mathematics ideas, I recommend some of Vy Hart's video doodles; you can find two at:

Spirals, Fibonacci, and Being a Plant (part 1) https://www.youtube.com/watch?v =ahXIMUkSXX0

Triangle Party https://www.youtube.com/watch?v=o6KlpIWhbcw&list=PLF7CBA45 AEBAD18B8&index=7

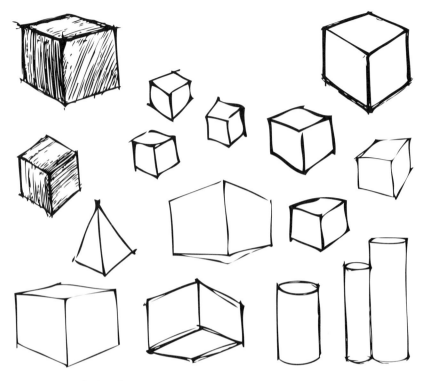

FIGURE 8.7   Math doodle

9. Students Write Questions and Tests

Ask students to design their own questions or write an assessment for other students. The act of writing a good question will help students focus on what is important and allow them to be creative in their thinking, which is itself important. Students really enjoy being given the task of writing a mathematics assessment.

Diagnostic Comments

All of the preceding strategies achieve the first two parts of the three-stage A4L process: they help students know what they are learning and where they should be in their learning.

Parts 1 and 2 of A4L are very important in their own right, but Part 3 gives students help that no other method can. This critically important part is for students to understand how to close the gap between where they are and where they need to be. In this realm there is one method that stands above all others in its effectiveness: the process of teachers giving students diagnostic comments on their work. One of the greatest gifts you can give to your students is your knowledge, ideas, and feedback on their mathematical development, when phrased positively and with growth messages.

Ellen Crews was one of the teachers I worked with in Vista School District. During our professional development sessions, I shared with Ellen and the other teachers the research I summarized earlier, showing the positive impact of diagnostic comments as a replacement to grades. Ellen is an incredible and dedicated teacher who works in a challenging school that is under "program improvement" (PI); that is, the state of California has deemed the school to be underperforming. The school has a diverse student intake, with 90% of students classified as Latino/a, and 10% a range of other ethnicities. Forty-three percent of students are English Language Learners, and 86% qualify for free school meals. Ellen told me that when she started teaching there the administrators were very focused on testing, as is typical with schools given a PI label. Test questions and chapter tests mirrored standardized assessment questions with multiple-choice answers. The school was focused on "improvement," which sounds good, and teachers spent hours poring over the reports generated by testing software to look for trends. Each student was then assigned a color, and the teachers were told that blue students would do fine on the state test, but red students did not stand much of a chance. The teachers were also told to focus most on the bottom end of the green students and the top end of the yellow students, as gains in their performance would have the greatest impact on school test scores.

Words like "improvement" sound laudable, but the method for improvement the school used, as is typical in PI schools, highlighted groups of students that would help school test scores, instead of the needs of individual students. The school first reduced students to test scores and then to colors, which determined which students could receive less attention, both high performing and low performing. This approach, of treating living children as statistics to be manipulated, is one used by schools across the United States, usually in response to harsh judgments and labels given in the name of "improvement."

In the midst of this performance culture imposed upon the school, Ellen decided to give a survey to her students. She found, unsurprisingly, that there were high levels of test anxiety. The students had taken numerous tests, and teachers had been told to continually stress the importance of successful performance. Ellen wanted to change this culture, and her first move was to stop giving chapter tests and to replace them with smaller assessments. She stopped using the words "test" and "quiz" and called her mini-assessments an opportunity to "Show What You Know." The multiple-choice questions were taken out, and students were asked to write answers to mathematics questions. Ellen also stopped preparing students for the district's benchmark exams; instead, she gave them without any warning, to avoid the buildup of anxiety, telling students, "Just do the best you can and don't worry about it." The students' scores on tests did not decline, despite Ellen's lack of preparation and focus upon them, and student anxiety decreased. Importantly, as Ellen told me, students started enjoying math class.

But Ellen—like other reflective, caring teachers—was not satisfied; she took further improvement steps. The next year, she and her eighth-grade teaching colleagues, Annette Wilson and Angela Townsend, stopped grading altogether, and instead used a rubric with mathematical statements. The whole team also changed the names of their assessments to "Show What You Can Do" (Exhibit 8.10). When the math team stopped giving scores and instead gave diagnostic

Show What You Can Do Self-Assessment

What we value from an individual	Justify (if necessary)	
Perseverance • Did you stick with it? • Did you try something else? • Did you ask a question? • Did you describe where you're stuck?		Did it! Approved
Multiple Representations Words Pictures Charts Diagrams Graphs More than one solution process Data Table		Did it! Approved
Clear Expectations • Did you describe your thinking process? • How did you get your answer? *or* Where did you get stuck? • Ideas: arrows, color, words, numbers		Did it! Approved
Product • Did you complete the task, or where did you get stuck? • Did you give the task your best effort?		Did it! Approved

Source: From Ellen Crews.

Exhibit 8.10

feedback, the students began to read and interpret the feedback, and occasionally to ask questions. Ellen made scores available to students if they asked. Ellen told me she found that at first she was needing to spend too much time on diagnostic feedback, which she was regularly giving to all of her 110 students, so she learned to write feedback when it was most helpful to students. This is the perfect approach to diagnostic feedback. It does take more time than a check or grade, but it is considerably more helpful to students. Occasional feedback, given at professionally judged times of importance, is an invaluable gift to students, but it does not have to be given often.

Ellen now tells me that many more of her students are putting effort into math, and she sees them strive hard to do the very best they can, which is an ideal result. Ellen also told me about the improvements to her teaching that are coming from the helpful information her new forms of assessment give her and that she uses to plan her teaching. In the years following the changes made in Ellen's teaching and assessing, in her middle school, the grades of the students moving to high school significantly improved, and failure rates in high school algebra halved.

Advice on Grading

Many teachers, unfortunately, are forced into grading, as it is a requirement of their school district or the administrators at their school. Ideally, teachers are asked to provide grades only at the end of a course—not during the course, when students need information only on ways to learn better, which should be given through formative assessment. The following list compiles advice on ways to grade fairly and to continue communicating positive growth messages even when faced with grading requirements:

1. *Always allow students to resubmit any work or test for a higher grade*—this is the ultimate growth mindset message, communicating to students that you care about *learning*, not just performance. Some teachers tell me this is unfair, as students may go away and learn on their own what they need to, to improve their grade; but we should value such efforts, as they are, at their core, about learning.

2. *Share grades with school administrators but not with the students.* If your school requires grades before the end of a course, this does not mean that you must give them to students. Instead, give students verbal or diagnostic written feedback on ways to improve.

3. *Use multidimensional grading.* Many teachers believe in the breadth of mathematics and may value multidimensional mathematics in the classroom, but assess students only on whether they get correct answers to procedural questions. The best teachers I have worked with who had to give grades have used students' mathematics work rather than test performance—recording, for example, whether they ask questions,

show mathematics in different ways, reason and justify, or build on each other's thinking. In other words, they assess the multidimensionality of math. When students are assessed on many ways of working in mathematics, many more students are successful.

4. *Do not use a 100-point scale.* One of the most unfair and mathematically egregious methods of grading is when teachers use a few assignments as the basis for a grade, assuming each is worth 100 points and giving zero points for any incomplete, missing, or failed assignment. Douglas Reeves (2006) has shown that such practices defy logic, as the gap between students receiving an A, B, C, or D is always 10 points, but the gap between a D and an F is 60 points. This means that one missing assignment could mean a student drops from achieving an A for a class to getting a D. Reeves's recommendation is to use a 4-point scale:

A = 4
B = 3
C = 2
D = 1
F = 0

in which all intervals are equal, rather than:

A = 91+
B = 81–90
C = 71–80
D = 61 -70
F = 0

which is a mathematically nonsensical scale (Reeves, 2006).

5. *Do not include early assignments from math class in the end-of-class grade.* When teachers do this, they are essentially grading students on their work from a previous class. Grades should show what students have learned in a class, not what they had learned from another class. Grading should include only work and assignments from the point in the class when students are working on what is learned in the class.

6. *Do not include homework, if given, as any part of grading.* As Chapter Six explained, homework is one of the most inequitable practices in education; its inclusion in grading adds stress to students and increases the chances of inequitable outcomes.

Conclusion

When teachers give assessments to students, they are faced with an incredible opportunity: to provide students with information about their learning—rather than their achievement—which accelerates pathways to success and gives students powerful growth mindset messages about mathematics and learning. We have considerable research evidence showing that a change from grading and testing to A4L methods has a powerful positive impact on students' achievement, self-beliefs, motivation, and future learning pathways. This chapter has shared some of the work of dedicated and insightful teachers who have made this shift. In the final chapter we will review all of the ways that growth mindset mathematics classrooms can be created and sustained.

CHAPTER 9

Teaching Mathematics for a Growth Mindset

My goal in writing this book has been to give math teachers, leaders, and parents a range of teaching ideas that will enable students to see mathematics as an open, growth, learning subject and themselves as powerful agents in the learning process. I realize in writing this, my last chapter, that we have been on quite a journey, moving all the way from the ways we think about children's potential to the forms of assessment that can create responsible, self-regulating learners. In this chapter I will provide a set of teaching ideas, drawing from throughout the book, that can help you create and maintain a growth mindset mathematics classroom. This chapter offers a shorter summary of many of the ideas from the book, pulled together to give a more concise guide to setting up a growth mindset mathematics class.

Encourage All Students

FIGURE 9.1   Inspiring all math learners

Setting Up Classroom Norms

Students come into class unsure of what expectations teachers will have for them. The first days of class and even the first hours of the first day are a great time to establish classroom norms. I often start my own classes just telling students what I do and do not value. I tell them that:

- I believe in every one of them, that there is no such thing as a math brain or a math gene, and that I expect all of them to achieve at the highest levels.

- I love mistakes. Every time they make a mistake their brain grows.

- Failure and struggle do not mean that they cannot do math—these are the most important parts of math and learning.

- I don't value students' working quickly; I value their working in depth, creating interesting pathways and representations.

- I love student questions and will put these onto posters that I hang on the walls for the whole class to think about.

But all of these statements are just words—they are important words, to be sure, but they will be worthless if the students do not see the words supported by their teachers' actions.

We shared on Youcubed the seven most important norms to encourage on the first days of class and throughout the year, and the chapters in this book have reviewed ways to establish each. Some teachers have found it helpful to put up our Youcubed poster on classroom walls at the start of class (see Exhibit 9.1 and Setting up Positive Norms in Math Class in Appendix B.).

POSITIVE NORMS TO ENCOURAGE IN MATH CLASS

1. **Everyone Can Learn Math to the Highest Levels.**

   Encourage students to believe in themselves. There is no such thing as a "math person." Everyone can reach the highest levels they want to, with hard work.

2. **Mistakes Are Valuable.**

   Mistakes grow your brain! It is good to struggle and make mistakes.

3. **Questions Are Really Important.**

   Always ask questions, always answer questions. Ask yourself: why does that make sense?

4. **Math Is about Creativity and Making Sense.**

   Math is a very creative subject that is, at its core, about visualizing patterns and creating solution paths that others can see, discuss, and critique.

5. **Math Is about Connections and Communicating.**

   Math is a connected subject, and a form of communication. Represent math in different forms—such as words, a picture, a graph, an equation—and link them. Color code!

6. **Depth Is Much More Important Than Speed.**

   Top mathematicians, such as Laurent Schwartz, think slowly and deeply.

7. **Math Class Is about Learning, Not Performing.**

   Math is a growth subject; it takes time to learn, and it is all about effort.

Exhibit 9.1

As well as telling students about norms and expectations, I find it valuable for students to communicate their own desired norms for working in groups together. Before any students work on math in a group with others, I ask them to discuss in small groups what they do and do not appreciate from other students, as I reviewed in Chapter Seven, and produce posters of these preferences. This is a worthwhile activity, as it helps students enact positive norms knowing they are shared by their peers, and teachers can refer students back to the posters later if good group work behavior needs to be reestablished.

The Railside teachers I discussed in Chapter Seven encouraged good group work very carefully, teaching students how to work well in groups—listening to each other, respecting each other, and building on each other's ideas. The teachers decided that in the first 10 weeks of high school they would not focus on the mathematics students learned but on group norms and ways of interacting. The students worked together on math all of the time, but the teachers did not worry about content coverage, only about students' learning of respectful group work. This careful teaching of good group work was reflected in the students' impressive mathematics achievements in the four years of high school (Boaler & Staples, 2005).

The Participation Quiz

My favorite strategy for encouraging good group work—a strategy that can be used early and often—is to ask students to take a participation quiz. The authors who conceptualized complex instruction (Cohen & Lotan, 2014) recommend that the participation quiz is graded. This does not involve grading individuals, which gives a negative fixed message; rather, it involves grading the behavior of groups. But the participation quiz doesn't have to end with a grade; it just needs to give students a strong message that it matters how they interact and that you are noticing. I really like this grouping strategy; I have taught it to groups of teachers who later told me that it quickly transformed the ways students worked in groups.

To run a participation quiz, choose a task for students to work on in groups, then show them the ways of working that you value. For example, the slides shown in Exhibits 9.2 and 9.3 come from the highly successful Railside teachers. In the first, the teachers highlight mathematical ways of working that they value. With younger children a much smaller list could be used.

The second focuses on the ways of interacting that lead to good group work.

These could be presented on posters in the room instead of on slides. Once you have shown these to students, you can start them working. As they work together in groups, walk around the room watching group behavior, writing down comments. To do this you will need a piece of paper, or an area on the white board, divided into sections, with a space for each group. For example, with 32 students working in 8 groups of 4:

1.	2.	3.	4.
5.	6.	7.	8.

Participation Quiz Mathematical Goals

Your group will be successful today if you are …

- Recognizing and describing patterns
- Justifying thinking and using multiple representations
- Making connections between different approaches and representations
- Using words, arrows, numbers, and color coding to communicate ideas clearly
- Explaining ideas clearly to team members and the teacher
- Asking questions to understand the thinking of other team members
- Asking questions that push the group to go deeper
- Organizing a presentation so that people outside the group can understand your group's thinking

No one is good at all of these things, but everyone is good at something. You will need all of your group members to be successful at today's task.

Source: From Carlos Cabana.

Exhibit 9.2

Participation Quiz Group Goals

During the participation quiz, I will be looking for …

- Leaning in and working in the middle of the table
- Equal air time
- Sticking together
- Listening to each other
- Asking each other lots of questions
- Following your team roles

Source: From Carlos Cabana.

Exhibit 9.3

As you circulate and take notes, quote students' actual words when they are noteworthy. Some teachers do this publicly, writing comments onto a white board at the front of the room. Others use paper clipped to a board. At the end of the lesson, you should have a completed chart and can assign the student groups a grade, or give students feedback on their group work without the grade. The following is an example of a teacher's grading of a participation quiz:

Quick start	All four working	"How do you guys think?"	Talking about clothes
All working together	Checking each other's work	Building shape in middle of table	Off task—group asked to stop
<u>Great</u> discussions	Asking nice questions: "How would that work with another number?"	Checking with each other	Individual working, no discussion
Staying together		A+	B
"Let's go round and find out how everyone sees the shape"	Good group roles		
A+	A+		
Trying ideas	Slow task—off task	"Can anyone see a different way?"	Started well
Asking questions of each other	Building shape in the middle	Nice explaining to each other	Quiet reading
Talking about work	Checking ideas	Great debate on meaning	All focused all the time
A	Good discussion	A+	Asking nice questions
	A		A+

The notes do not have to be detailed, but they will help students understand what you value and will help them become much more attentive to the ways they interact with other students. My Stanford students and many teachers whom I have taught this method enjoy participation quizzes, comically "leaning in" and asking deep questions when I linger by their tables with my notepad! Students also have fun, at the same time realizing much more clearly what they need to do to engage well in a group.

I am a great believer in the participation quiz. Teachers who have used it in classes in which they previously had a lot of trouble encouraging good group work have been stunned at the positive change in students. Almost overnight, students start asking each other good questions and thinking about the equal involvement of different team members. When students are working well in groups, respecting each other and asking good questions, then classrooms are a great place to be in, for students and teachers.

Believe in All of Your Students

I have always known how important it is that students know their teacher believes in them; I knew this as a teacher and more recently became more acutely aware of it as a parent. When my daughter was five, she realized the teacher of her class in England was giving other students harder math problems, and she came home to me and asked why. When she realized that the teacher did not think she had potential—and sadly, this was true; the teacher had decided she had limited ability—her self-belief was shattered, and she developed a terribly fixed mindset that damaged her learning and confidence for a long time afterward. Now, some years later, after a lot of work from her parents and some wonderful teachers, she is transformed: she has a growth mindset and loves math. Despite the fact that the teacher never said to my daughter that she did not believe in her, she managed to communicate that message loud and clear, and this was understood by my daughter even at the young age of five.

The school that my daughter attended in England put students into ability groups in second grade, but they stopped this practice after reading the research evidence and learning about the strategies for teaching heterogeneous groups. After they made this change, the principal wrote to tell me it had transformed math classes and raised achievement across the school. If students are placed into ability groups, even if they have innocuous names such as the red and blue groups, students will know, and their mindsets will become more fixed. When children were put into ability groups in my daughter's school, children from the lower groups came home saying "All the clever children have gone into another group now." The messages the students received about their potential as learners in general (not just about math) were devastating for them. One of the first steps we need to take, as a nation, is to move away from outdated methods of fixed mindset grouping and communicate to all students that they can achieve.

The importance of students thinking their teacher believes in them was confirmed in a recent study that had an extremely powerful result (Cohen & Garcia, 2014). Hundreds of students were involved in this experimental study of high school English classes. All of the students wrote essays and received critical diagnostic feedback from their teachers, but half the students received a single extra sentence on the bottom of the feedback. The students who received the extra sentence achieved at significantly higher levels a year later, even though the teachers did not know who received the sentence and there were no other differences between the groups. It may seem incredible that one sentence could change students' learning trajectories to the extent that they achieve at higher levels a year later, with no other change, but this was the extra sentence:

"I am giving you this feedback because I believe in you."

Students who received this sentence scored at higher levels a year later. This effect was particularly significant for students of color, who often feel less valued by their teachers (Cohen & Garcia, 2014). I share this finding with teachers frequently, and they always fully understand its significance. I do *not* share the result in the hope that teachers will add this same sentence to all of their students' work. That would lead students to think the sentence was not genuine, which would be counterproductive. I share it to emphasize the power of teachers' words and the beliefs they hold about students, and to encourage teachers to instill positive belief messages at all times.

Teachers can communicate positive expectations to students by using encouraging words, and it is easy to do this with students who appear motivated, who learn easily, or who are quick. But it is even more important to communicate positive beliefs and expectations to students who are slow, appear unmotivated, or struggle. It is also important to realize that the speed at which students appear to grasp concepts is not indicative of their mathematics potential (Schwartz, 2001). As hard as it is, it is important to not have any preconceptions about who will work well on a math task in advance of their getting the task. We must be open at all times to any student's working really well. Some students give the impression that math is a constant struggle for them, and they may ask a lot of questions or keep saying they are stuck, but they are just hiding their mathematics potential and are likely to be suffering from a fixed mindset. Some students have had bad math experiences and messages from a young age, or have not received opportunities for brain growth and learning that other students have, so they are at lower levels than other students, but this does not mean they cannot take off with good mathematics teaching, positive messages, and, perhaps most important, high expectations from their teacher. You can be the person who turns things around for them and liberates their learning path. It usually takes just one person—a person whom students will never forget.

Value Struggle and Failure

Teachers care about their students and want them to do well, and they know that it is important for students to feel good about math. Perhaps it is this understanding that has led to most math classrooms across the United States being set up so that students get most of their work correct. But the new brain evidence tells us this is not what students need. The most productive classrooms are those in which students work on complex problems, are encouraged to take risks, and can struggle and fail and still feel good about working on hard problems. This means that mathematics tasks should be difficult for students in order to give students opportunities for brain growth and making connections, but it doesn't mean just increasing the difficulty, which would leave students frustrated. Rather, it means changing the nature of tasks in math classrooms—giving more low floor, high ceiling tasks. As discussed in Chapter Five, having a low floor means that anyone can access the ideas. Having a high ceiling means that students can take the ideas to high levels.

As well as changing tasks, teachers should communicate frequently that struggle and failure are good. Many of the students I teach at Stanford have achieved at high levels all their lives and received a lot of damaging fixed mindset feedback, being told that they are "smart" at frequent intervals. When they encounter harder work at Stanford and don't receive an A for everything, some of them fall apart, feeling devastated and questioning their ability. When they work on material that causes them to struggle—actually a very worthwhile place to be for learning—they quickly lose confidence and start to doubt that they are "smart" enough to be at Stanford. These are the students who have been brought up in a performance culture, for whom struggle and failure have never been valued. The students in my freshman class tell me how important the ideas we learn have been to them, how learning that struggle is good has kept them in math and engineering classes and has stopped them dropping out of STEM pathways.

We must work hard to break the myth of "effortless achievement," pointing out that all high achievers have worked hard and failed often, even those thought of as "geniuses," as Chapter Four discussed. We must also resist valuing "effortless achievement"—praising students who are fast with math. Instead, we should value persistence and hard thinking. When students fail and struggle it does not mean anything about their math potential; it means that their brains are growing, synapses are firing, and new pathways are being developed that will make them stronger in the future.

Give Growth Praise and Help

When Carol Dweck worked with preschool children, she found that some children were persistent when they experienced failure and wanted to keep trying, whereas others would give up easily and request that they repeat tasks that were easy for them. These persistent and nonpersistent mindset strategies were evident in children who were only three and four years old.

When researchers then conducted role-plays with the children and asked them to pretend to be an adult responding to their work, the persistent children role-played adults focusing on strategies, saying that the children would be more successful with more time or a different approach. The nonpersistent children role-played an adult saying that the child could not finish the work and so should sit in their room. The nonpersistent children seemed to have received feedback that told them they had personal limitations and that failure was a bad thing (Gunderson et al., 2013). This study, as well as many other mindset studies (Dweck, 2006a, 2006b; Good, Rattan, & Dweck, 2012), tell us that the forms of feedback and praise we give students are extremely important. We know that one way we aid and abet students in developing a fixed mindset is by giving them fixed praise—telling them, in particular, that they are smart. When students hear that they are smart, they feel good at first, but when they struggle and fail—and everyone does—they start to believe they are not so smart. They continually judge themselves against a fixed scale of "smartness," and this will be damaging for them, even if they get a lot of positive smart feedback, as the case of the Stanford students illustrates.

Instead of telling children that they are smart or clever, teachers and parents should focus on the particular strategies children have used. Instead of saying "You are so smart," it is fine to say to students something like "It's great that you have learned that," or "I love how you are thinking about the problem." Removing the word "smart" from our vocabulary is difficult to do, as we are all used to referring to people as smart. My undergraduates have really worked on this and now praise people for having good thinking and for being accomplished, learned, hard working, and persistent.

When students get work wrong, instead of saying "That is wrong," look for their thinking and work with it. For example, if students have added $\frac{1}{3}$ and $\frac{1}{4}$ and decided the answer is $\frac{2}{7}$, you could say: "Oh, I see what you are doing; you are using what we know about adding whole numbers to add the top and bottom numbers, but these are fractions, and when we add them we have to think about the whole fraction, not the individual numbers that make up the fraction." There is always some logic in students' thinking, and it is good to find it, not so that we avoid the "failure" idea, but so that we honor students' thinking. Even if children have completed a task that is completely wrong, be careful not to give the idea that the task is too hard for them, as this gives the idea that their ability is limited. Instead focus on strategies, saying such things as "You haven't learned the strategies you need for that yet, but you will soon."

It is important not to provide too much help to students and take away from the cognitive demand of tasks. Guy Brousseau, a French researcher, identified what he called "the didactic contract," which has since been recognized by teachers and researchers worldwide (Brousseau, 1984; Brousseau, 1997). Brousseau describes a common situation in mathematics classes whereby teachers are called over to students who ask for help; the students expect to be helped, and teachers know it is their role to help them, so the teachers break down the problem and make it easier. In doing so they empty the problem of its cognitive demand. Brousseau points out that this is a shared action between teachers and students; they are both playing the roles expected of them, fulfilling the "didactic contract" that has been established in classrooms, which results in students missing the opportunity to learn. Under the contract, students expect to not be allowed to struggle; they expect to be helped, and teachers know *their* role is to help students, so they jump in and help them, often unknowingly robbing them of learning opportunities. Textbook authors are complicit in a similar process, breaking problems down into small parts for students to answer. When my students ask for help, I am very careful not to do the mathematical thinking for them. Instead, I often ask students to draw the problem, which invariably unlocks new ideas for them.

I recently read about a second-grade teacher, Nadia Boria, who offers this response to students when they ask for help: "Let's think about this for a minute. Do you want my brain to grow or do you want to grow your brain today?" (Frazier, 2015).

This is a lovely response, and although teachers have to judge every interaction with their professional knowledge and intuition, knowing when students can handle more struggle and not get discouraged, it is important to remember that *not* helping students is often the best help we can give them.

The norms we set up for students in our math classes, the ways we help and encourage them, and the messages we give them are extremely important, but I cannot emphasize too strongly that giving students growth mindset messages will not help them unless we also show them that math is a growth subject. The remainder of this chapter will focus on the strategies and methods teachers can use to teach students open, growth, creative mathematics.

Opening Mathematics

Teach Mathematics as an Open, Growth, Learning Subject

The majority of mathematics questions that are used in math classrooms and homes are narrow and procedural and require students to perform a calculation. When students spend most of their math time working in this way, it is very hard for them to truly believe that math is a growth subject, as the closed questions communicate the idea that math is a fixed, right-or-wrong subject. It is reasonable for some questions to be narrow, with one right answer, but such questions are not necessary for students to develop a sound mathematical understanding, and they should be the minority of questions, if used at all. Mathematics tasks should offer plenty of space for learning. Instead of requiring that students simply give an answer, they should give students the opportunity to explore, create, and grow.

Any math task can be opened up, and when they *are* opened, many more students engage and learn. Here are four examples of ways to open math tasks:

1. Instead of asking students to answer the question 1/2 divided by 1/4, ask them to make a conjecture about the answer to 1/2 divided by 1/4 and make sense of their answer, including a visual representation of the solution. As I described in Chapter Five, when Cathy Humphreys asked students to solve $1 \div \frac{2}{3}$ she started by saying, "You may know a rule for solving this question, but the rule doesn't matter today, I want you to make sense of your answer, to explain why your solution *makes sense*."

2. Instead of asking students to simplify $\frac{1}{3}(2x + 15) + 8$, a common problem given in algebra class, ask students to find all the ways they can represent $\frac{1}{3}(2x + 15) + 8$ that are equivalent. Figure 9.2 shows examples.

3. Instead of asking students how many squares are in the 100th case, ask them how they see the pattern growing, and to use that understanding to generalize to the 100th case (see Figure 9.3).

Any math task can be opened up so that it offers students space for learning, as Chapter Five discussed more fully. For example, you can ask students to discuss:

$\frac{1}{3}(2x + 1S) + 8$	$\frac{2x + 1S}{3} + 8$	$\frac{2}{3}x + S + 8$
$\frac{2x}{3} + 13$	$\frac{2x + 1S + 24}{3}$	$\frac{1}{3}(2x + 39)$

FIGURE 9.2   Algebra examples

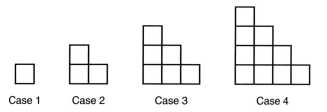

FIGURE 9.3 Stairs

- Ways of seeing the mathematics

- Ways of representing ideas

- The different pathways through the problem and strategies

- The different methods used: "Why did you choose those methods? How do they work?"

When students are working on open math tasks, they are not only encouraged to see mathematics as a growth subject but also placed in the role of an inquirer. They are no longer finding an answer; they are exploring ideas, making connections, and valuing growth and learning. At the same time they are making these inquiries, they are learning formal mathematics—the methods and formulas set out in curriculum standards. The difference is, students learn standard methods when they encounter a need for them, which gives them motivation and excitement to learn them (Schwartz & Bransford, 1998). As I've emphasized, the most perfect open mathematics questions are those with a low floor and a high ceiling (see the "Tasks" collection on the Youcubed website, http://www.youcubed.org/tasks/). When considering whether a task is open, the most important question to ask, in my view, is whether it offers space for students to learn.

Encourage Students to Be Mathematicians

Mathematicians see their subject as creative, beautiful, and aesthetic. All children can work as mathematicians, and encouraging them to be mini mathematicians can be very empowering. It is important that students engage in proposing ideas—or, to use the mathematical term, making conjectures about mathematics. One of the most amazing third-grade teachers I have ever witnessed is Deborah Ball, now the dean of the school of education at the University of Michigan. Deborah taught her third-grade students to be mathematicians—to be inquirers, and to make conjectures about math. When the class came to a consensus on a mathematical idea, they would say they had a "working definition," which they would return to and refine when they had explored further. In one class period, third-grader Sean made a proposal about the number 6, saying that it can be both even or odd (video is available online: Mathematics Teaching and Learning to Teach, 2010; http://deepblue.lib.umich.edu/handle/2027.42/65013).

His reason for this conjecture was that 6 is made up of an odd number of groups of 2, whereas other even numbers, such as 4 and 8, have even numbers of groups of 2. Many students in the class argued with Sean, returning to the class's working definition of an even number. Most teachers would have told Sean that he was incorrect and moved on, but Deborah was interested in his thinking. What followed was an animated discussion among the students that has captured viewers of many backgrounds and persuasions, including teachers and mathematicians. In the lesson the children became deeply involved in thinking abut Sean's conjecture, and at no point did they ask the teacher to tell them if Sean was right or not, which would have shut down the conversation. Instead, the third graders asked Sean to prove his conjecture, and they offered counter evidence, using many different definitions of an even number to show Sean that 6 was even and not also odd. At some point in the discussion Deborah realized that Sean had proposed something about the number 6, and other numbers that share the quality of having an odd number of multiples of 2, such as 10, that does not have a name in mathematics, and the class decided to call the numbers "Sean numbers." Sean was making an observation that was not wrong; he was pointing out that some numbers have different characteristics. In later discussions in the year the class would explore numbers and casually make reference to the use of "Sean numbers" when they came up. Unlike many third graders, who are turned off math by its procedural presentation, these children loved being able to share their thinking and ideas, and to make conjectures and construct proofs as they came to agree, as a class, on working definitions and propositions, at the same time they were learning formal mathematics. They students were excited to work on conjectures, reasoning, and proof, and they looked to any observer like young mathematicians at work (Ball, 1993).

Some people are shocked by the idea of calling children mathematicians, yet they comfortably refer to children as young artists and scientists. This is because of the pedestal that mathematics is placed on, as I discussed in Chapter Six. We need to counter the idea that only those with many years of graduate mathematics should act as mathematicians. We need to stop leaving the experience of real mathematics to the end, when students are in graduate school, by which time most students have given up on math. There is no better way to communicate to all students that mathematics is a broad, inquiry-based subject that they can all work on than by asking children to be mathematicians.

Teach Mathematics as a Subject of Patterns and Connections

Mathematics is all about the study of patterns. Many people appreciate that they are working with patterns when working on problems such as the one shown in Figure 9.4, where they are asked to extend the pattern.

But even when learning arithmetic, or more abstract areas of math, the work of any student is about seeking patterns. I have tried to encourage my own children to see themselves as pattern

FIGURE 9.4   Pattern strip

seekers, and I was pleased recently when my eight-year-old daughter was working on division. She had just been taught the "traditional algorithm" for division, but when she was given questions like this …

$$6\overline{)18} \qquad 7\overline{)35} \qquad 8\overline{)27}$$

$$8\overline{)96} \qquad 6\overline{)72} \qquad 7\overline{)83}$$

… she found that the algorithm was useful only in some cases. After working on some questions, she said—"Oh, I can see a pattern; the dividing loop method" [by which she meant the "traditional algorithm"] "only helps when the first digit is bigger than the number being divided." I am not a big fan of students learning division through the traditional algorithm, as it often prevents students from looking at the whole number and works against place value understanding, but I was pleased that her pattern-seeking orientation meant she was thinking about patterns in the numbers and not blindly following a method. I am not suggesting that the traditional algorithm is not useful; it may be helpful *after* students understand division as one of many division strategies. When they are learning division, students should use methods that encourage their understanding of the numbers involved and the concept of division.

When teachers teach mathematical methods, they are really teaching a pattern—they are showing something that happens all of the time, something that is *general*. When we multiply a number bigger than 1 by 10, then the answer will have a zero. When we divide the circumference of a circle by twice its radius, we always get the number pi. These are patterns, and when students are asked to see mathematics as patterns, rather than methods and rules, they become excited about mathematics. They can also be encouraged to think about the nature of the patterns—what is general about the case? Keith Devlin, a top mathematician and NPR's "math guy," has written a range of excellent books for the public. In one of my favorites, *Mathematics: The Science of Patterns*, Devlin shows the work of mathematicians as being all about the use and study of patterns—arising from what he describes as the natural world or the human mind. Devlin quotes the great mathematician W. W. Sawyer saying that "mathematics is the classification and study of all possible patterns" and that patterns include "any kind of regularity that can be recognized by the mind." Devlin agrees, saying "Mathematics is not about numbers, but about life. It is about the world in which we live. It is about ideas. And far from being dull and sterile as it is so often portrayed, it is full of creativity" (Devlin, 2001).

Invite students into the world of pattern seeking; give them an active role in looking for patterns in all areas and all levels of math.

In Chapter Three I introduced Maryam Mirzakhani, a mathematician and a colleague of mine at Stanford. She hit the news headlines worldwide when she became the first woman to win the Fields Medal. As mathematicians talked about the tremendous contributions she has made to the advancement of mathematics, they talked about the ways her work connects many areas of mathematics, including differential geometry, complex analysis, and dynamical systems. Maryam reflected: "I like crossing the imaginary boundaries people set up between different fields—it's very refreshing … there are lots of tools, and you don't know which one would work. It's about being optimistic and trying to connect things." This is a mindset I would love all students of mathematics to have.

When students make and see connections between methods, they start to understand real mathematics, and they enjoy the subject much more. This is a particular imperative for getting more girls into STEM fields, as discussed in Chapter Six. Curriculum standards often work against connection making, as they present mathematics as a list of disconnected topics. But teachers can and should restore the connections by always talking about and valuing them and asking students to think about and discuss connections. The mathematical connections video we provide on Youcubed shows the ways fractions, graphs, triangles, rates, Pythagoras' theory, tables, graphs, shapes, slope, and multiplication are all connected under the theme of proportional reasoning (Youcubed at Stanford University, 2015c; http://www.youcubed.org/tour-of-mathematical-connections/). We made this video to show the connections between mathematical areas that students may not think exist, and teachers have found it helpful to show it to their students, to help them think about connections. From there students should be encouraged to explore and see mathematical connections in many different ways.

Here are some examples of ways to highlight connections in mathematics.

- Encourage students to propose different methods to solve problems and then ask them to draw connections between methods, discussing for example, how they are similar and different or why one method may be used and not another. This could be done with methods used to solve number problems, such as those shown in Figure 5.1, in Chapter Five.

- Ask students to draw connections between concepts in mathematics when working on problems. For example, consider the two mathematics problems in Exhibit 9.4 and Figure 9.5.

Teachers can encourage students to produce more than one representation and to connect the numbers in their solutions to their diagrams, allowing for the use of different brain pathways.

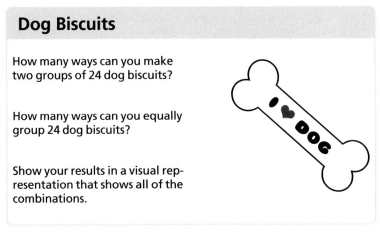

Dog Biscuits

How many ways can you make two groups of 24 dog biscuits?

How many ways can you equally group 24 dog biscuits?

Show your results in a visual representation that shows all of the combinations.

Exhibit 9.4

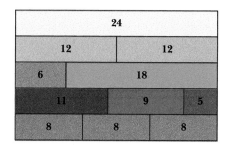

FIGURE 9.5 Dog biscuits solution

Some students might like to use grid paper, others a number line, others may use multilink cubes or other small objects. Teachers can ask students to think about different methods that may be used when considering equal grouping—in particular, addition and multiplication—and to think about how they are related.

In the different activities in Exhibit 9.5, students are asked to specifically focus on different areas of mathematics and the connections between them. Successful students are not those who think of mathematics as a series of disconnected topics—a view held by many students. Rather, they are the students who see mathematics as a set of connected ideas (Program for International Student Assessment [PISA], 2012), a viewpoint that teachers need to actively encourage, especially if textbooks give the opposite impression. Connected mathematics is inspiring and appealing to students, and all teachers can enable students to see the connected nature of mathematics.

Teach Creative and Visual Mathematics

In my own teaching of mathematics, I encourage student creativity by posing interesting challenges and valuing students' thinking. I tell students I am not concerned about their finishing math problems quickly; what I really like to see is an interesting representation of ideas, or a creative method or solution. When I introduce mathematics to students in this way, they always surprise me with their creative thinking.

It is very important to engage students in thinking visually about mathematics, as this gives access to understanding and to the use of different brain pathways. Amanda Koonlaba, a fourth-grade teacher who connects art to core school subjects, including mathematics, describes a time when she asked her students what kinds of arts lessons they had enjoyed in their core subjects. She recalls that one student "spoke quietly but enthusiastically, explaining that he loves visual art because creating helps him 'forget the bad' and he needs that 'more than once a week'" (Koonlaba, 2015).

Art and visual representations don't play only a therapeutic and creative role, although both are important. They also play a critical role in opening access to understanding for all students. When I ask students to visualize and draw ideas, I always find higher levels of engagement and opportunities to understand the mathematical ideas that are not present without the visuals.

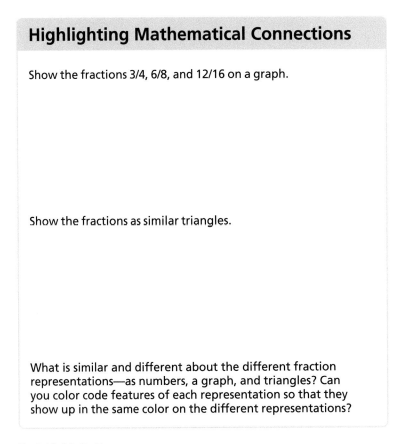

Highlighting Mathematical Connections

Show the fractions 3/4, 6/8, and 12/16 on a graph.

Show the fractions as similar triangles.

What is similar and different about the different fraction representations—as numbers, a graph, and triangles? Can you color code features of each representation so that they show up in the same color on the different representations?

Exhibit 9.5

Some students find visual ideas harder than others, but those are the students who will be most helped by using them.

As well as asking students to draw ideas, methods, solutions, and problems, teachers should always ask them to connect visual ideas with numerical or algebraic methods and solutions. Color coding, as I showed in Chapter Five, is a good way to encourage these connections. In the next two examples we see how much color can enhance students' understanding of geometry, fractions, and division; in earlier chapters I showed color coding in algebra and parallel lines. When students learn about angle relationships, they can be asked to color the different angles of a triangle, tear them off, and line them up to see the angle relationship. The visual depiction of the angles will help them remember the relationships.

Understanding of fractions also can be enhanced if students are asked to color code the fractions (see Exhibit 9.6 and Figure 9.6).

I particularly like a color coding approach to division created by Tina Lupton, Sarah Pratt, and Kerri Richardson. They propose that students be asked to solve division problems using a

Color Coding Brownies

Sam has made a pan of brownies that he wants to cut into 24 equal pieces. He wants to share them equally with 5 of his friends. Partition the pan of brownies and use color coding to show how many Sam and his friends will get.

<table>
<tr><td></td><td></td><td></td><td></td><td></td></tr>
<tr><td></td><td></td><td></td><td></td><td></td></tr>
<tr><td></td><td></td><td></td><td></td><td></td></tr>
<tr><td></td><td></td><td></td><td></td><td></td></tr>
</table>

Exhibit 9.6

Sam has made a pan of brownies that he wants to cut into 24 equal pieces. He wants to share them equally with 5 of his friends. Partition the pan of brownies and use color coding to show how many Sam and his friends will get.

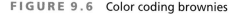

FIGURE 9.6   Color coding brownies

division quilt, which helps students really see and understand the partitioning of numbers into equal groups and remainders (see Figure 9.7). They give more details on ways to structure this helpful activity in Lupton, Pratt, and Richardson (2014).

Representing mathematical ideas in different ways is an important mathematical practice, used by mathematicians and high-level problem solvers. When mathematicians work, they represent ideas in many different ways—with graphs, tables, words, expressions, and—less well

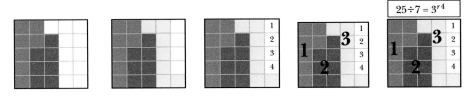

FIGURE 9.7   Division quilts
Source: Lupton, Pratt, & Richardson, 2014.

known—drawings and even doodles. Mirzakhani describes thinking about a difficult math problem:

> "You don't want to write down all the details … But the process of drawing something helps you somehow to stay connected." Mirzakhani said that her 3-year-old daughter, Anahita, often exclaims, "Oh, Mommy is painting again!" when she sees the mathematician drawing. "Maybe she thinks I'm a painter." (Klarreich, 2014)

Whenever I am given a complex mathematics problem to solve, I draw it; it is the best way I know for tackling a hard problem and understanding the mathematics. When I work with students and they are stuck, I also ask them to draw, asking questions such as "Have you tried drawing the problem?" Students who are not used to drawing mathematics may find this a challenge at first, but they can learn to draw, and it will help them. Chapter Eight gives more ideas on ways to engage students in drawing and doodling.

Being willing and able to use representations in mathematical thinking is immensely helpful for students, both in school mathematical work and in life.

Encourage Intuition and Freedom of Thought

I discussed in Chapter Five the ways in which high-level mathematics users—such as Sebastian Thrun, making robots for the Smithsonian—use intuition to develop mathematical ideas. Leone Burton interviewed 70 research mathematicians to find out about the nature of their work; 58 of them talked about the important role of intuition in their work. Reuben Hersh, in his book *What Is Mathematics Really?*, says that if we "look at mathematical practice, the intuitive is everywhere" (Hersh, 1999).

But what is this thing called intuition? And why are mathematicians engaging in intuition when students rarely if ever do so in classrooms? Teachers can encourage students to use intuition with any math problem simply by asking them what they think would work, before they are taught a method. Opportunities for students to think intuitively occur all through mathematics at every grade level. Elementary teachers could ask students before they teach any method, to work out their own method for solving the problem; for example, how they might find the area of a rug before being given an area formula. In middle or high school we can ask students how they might find the height of an object that is too tall to measure before we teach them methods to do so (see Boaler, Meyer, Selling, & Sun, n.d.). In Chapter Five I talked about the precalculus lesson in which students were asked to conjecture and think intuitively about finding the volume of a lemon, before they were taught calculus. Giving students the opportunity to use intuition is something that teachers can do with a small change in their practice.

When students are asked to use intuition to think about a mathematical idea, they are being invited to think openly and freely. When I asked a group of third-grade children who had learned through number talks what they thought about the number talks, the first thing Dylan said to me was, "You are free, you can do whatever you want. You can take numbers and break them down. . . ." Delia, one of the students featured in *Beyond Measure* (the second documentary from the director of *Race to Nowhere*), spoke in a similar way about her mathematical experiences after she was engaged in inquiry math: "I have a connection with math now. It's like, I'm open, I feel alive, I feel more energetic." In the same film, Niko compares his prior mathematics experience of working through worksheets with the collaborative, inquiry-based teaching he was then experiencing: "Math class last year, you were like, by yourself, every man for themselves, but this year, it's open, it's like a city, we are all working together to create this new beautiful world."

I continue to be amazed and inspired by the words children use to talk about mathematics when it is opened up, when they are asked to use their ideas and experience creative, beautiful mathematics. The fact that they say "We are free," "I am open, I feel alive," and "We are working together to create this new beautiful world" speaks to the transformative effect that inquiry-based mathematics can have. Students speak in these ways because they have been given intellectual freedom, and that is a very powerful and moving experience. When we ask students to use intuition and think freely, they develop not only a new perspective on mathematics, themselves, and the world but also an intellectual freedom that transforms their relationship with learning.

Deborah Ball has written an engaging and provocative paper in which she quotes from the legendary psychologist Jerome Bruner:

> "We begin with the hypothesis that any subject can be taught effectively in some intellectually honest form to any child at any stage of development. It is a bold hypothesis and an essential one in thinking about the nature of a curriculum. No evidence exists to contradict it; considerable evidence is being amassed that supports it." (Bruner, 1960, cited in Ball, 1993)

This statement can be challenging for many, and it was perturbing for my Stanford students when I first introduced the idea. But they willingly thought about the ways that ideas in calculus can be discussed with young children. Deborah Ball is quite convinced; she says that "the things that children wonder about, think and invent are deep and tough" (Ball, 1993, p. 374). If we release teachers and students from the prescribed hierarchy of mathematics given in content standards and allow students to explore higher-level ideas that could be very engaging—such as the fourth dimension, negative space, calculus, or fractals—then we have the opportunity to introduce them to real mathematical excitement and explore powerful ideas, at any age. I am not suggesting that we teach formal higher-level mathematics to young children, but I like the possibility that Bruner and Ball discuss—that any part of math can be introduced in an intellectually honest form at any age. This is an exciting and important idea.

Value Depth over Speed

One thing we need to change in mathematics classrooms around the world is the idea that in mathematics speed is more important than depth. Mathematics, more than any other subject, suffers from this idea, and the learners of mathematics suffer because of it. Yet our world's top mathematicians—people such as Maryam Mirzakhani, Steven Strogatz, Keith Devlin, and Laurent Schwartz, who all have won the highest honors for their work—all talk about working slowly and deeply and not being fast. I quoted from Laurent Schwartz in Chapter Four; this sentence comes from his longer quote: "What is important is to deeply understand things and their relations to each other." Schwartz talks about feeling "stupid" in school because he was a slower thinker, and he urges his readers to appreciate that mathematics is about depth and connections, not shallow knowledge of facts and fast work.

Mathematics is a subject that should be highlighting depth of thinking and relationships at all times. In a recent visit to China, I was able to watch a number of middle and high school math lessons in different schools. China outperforms the rest of the world on PISA and other tests, by a considerable margin (PISA, 2012). This leads people to think that math lessons in China are focused upon speed and drill. But my classroom observations revealed something very different. In every lesson I observed, teachers and students worked on no more than three questions in an hour's lesson. The teachers taught ideas—even ideas that are among the more definitional and formulaic in mathematics, such as the definitions of complementary and supplementary angles—through an inquiry orientation. In one lesson, the teacher explored the meaning of complementary and supplementary angles with students by giving an example and asking them to "ponder the question carefully" and then discuss questions and ideas that came up (video, Youcubed at Stanford University, 2015d; www.youcubed.org/high-quality-teaching-examples/). The ensuing discussion of complementary and supplementary angles moved into a depth of terrain I have never before seen in my observations of mathematics classrooms teaching this topic. The teacher provocatively took the students' ideas and made incorrect statements for the students to challenge, and the class considered together all of the possible relationships of angles that preserve the definitions.

The following extract is a transcript from a typical U.S. lesson on complementary and supplementary angles, taken from a TIMSS video study of teaching in different countries (Stigler & Hiebert, 1999):

Teacher: Here we have vertical angles and supplementary angles. Angle A is vertical to which angle?
Students chorus: 70 degrees.
Teacher: Therefore angle A must be?
Students chorus: 70 degrees.
Teacher: Now you have supplementary angles. What angle is supplementary to angle A?
Students Chorus: B
Teacher: B is, and so is … ?
Students: C.
Teacher: Supplementary angles add up to what number?
Students: 180 degrees.

In the extract we observe definitional questions with one answer, toward which the teacher is leading the students. Compare this with a lesson we watched in China, in which the teacher did not ask questions such as "Supplementary angles add up to what number?" Rather, she asked questions such as "Can two acute angles be supplementary angles? Can a pair of supplementary angles be acute angles?" These are questions that require students to think more deeply about definitions and relationships. Here is an extract from the lesson in China that I watched and that stands as an important contrast to the U.S. lesson.

Student: As he just said, if there are two equal angles, whose measures add up to 180 degrees, they must be two right angles. Because the measures of acute angles are always smaller than 90 degrees, the sum of the measures of two acute angles will not be larger than 180 degrees.
Teacher: Therefore, if two angles are supplementary, they must be two obtuse angles?
Student: That is not correct.
Teacher: No? Why?
Teacher: I think if two angles are supplementary, they must be two obtuse angles.
Student: I think they could be an acute angle and an obtuse angle.
Teacher: She says, although they cannot both be acute angles, they can be one acute angle and one obtuse angle.
Student: For example, just like the Angle 1 and Angle 5 in that question. One angle is an acute angle. The other one is an obtuse angle.
Teacher: OK. If two angles are supplementary, they must be one acute angle and one obtuse angle?
Student: That's still not accurate.
Student: You should say, if two angles are supplementary, at least one of them is an acute angle.
Other Students: No, at least one angle is larger than 90 degrees.
Student: An exception is when the two angles are right angles.

The lessons from the United States and from China could not have been more different. In the U.S. lesson, the teacher fired procedural questions at the students and they responded with the single possible answer. The teacher asked questions that could have come straight from books, that highlighted an easy example of the angle, and students responded with definitions they had learned. In the lesson in China, the teacher did not ask complete-this-sentence questions; she listened to students' ideas and made provocative statements in relation to their ideas that pushed forward their understanding. Her statements caused the students to respond with conjectures and reasons, thinking about the relationships between different angles.

The second half of the lesson focused on the different diagrams students could draw that would illustrate and maintain the angle relationships they had discussed. This involved the students producing different visual diagrams, flipping and rotating rays and triangle sides. Students discussed ideas with each other and the teacher, asking questions about the ideas, pushing them to a breadth and depth I had not imagined before seeing the lesson. As the class discussed the visual diagrams of angle relationships, one student reflected: "This is fascinating." There are not many students who would have drawn this conclusion from the U.S. version of the lesson.

The TIMSS video study compared teaching in the United States with teaching in other countries and concluded that U.S. lessons were "a mile wide and an inch deep" (Schmidt et al., 2002), whereas lessons in other countries they studied, particularly Japan, were conceptual and deeper, and involved more student discussion. The analysts linked the depth of the discussions and work in Japan, compared to the United States, as the reason for the higher achievement in Japan (Schmidt et al., 2002; Schmidt, McKnight, & Raizen, 1997).

Some parents' lack of understanding of the importance of mathematical depth, along with mis-guided beliefs that their children will be advantaged if they go faster, leads them to campaign for their children to skip grades and be taught higher-level mathematics as early as possible. But mathematics learning is not a race, and it is mathematical depth that inspires students and keeps them engaged and learning mathematics well, setting them up for high-level learning in the future. We know that students who are pushed to go through content faster are usually the ones to drop out of mathematics when they get the opportunity (Jacob, 2015; also Boaler, 2015b). We want all students to be productively engaged with math, and no student should find it too easy or be made to repeat ideas they have already learned. One of the best and most important ways to encourage high achievers is to give them opportunities to take ideas to greater depth, which they can do alongside other students, who may take ideas in depth on other days. A method I use to do this with my Stanford students is to ask those who finish problems to extend them, taking them in new directions.

Last week I gave my Stanford students a problem called "a painted cube," along with boxes of sugar cubes so that they could model the problem (see Exhibit 9.7 and Figure 9.8).

Some of the students built a smaller case, such as a $3 \times 3 \times 3$ cube, out of their sugar cubes and colored the outside surface with pens, to consider the way the cube sides were distributed.

I told them that when they had solved the 5×5 cube problem they could extend the problem in any way they wished. This was the best part of the lesson and the occasion for many more learning opportunities, as different groups considered, for example, how to work out the answer with a pyramid of cubes instead of a cube of cubes (Figure 9.9); another group worked out the relationships in a pyramid made of smaller pyramids, and still another worked out relationships if the cube moved into the 4th dimension and from there n-dimensions.

If you give students the opportunity to extend problems, they will almost always come up with creative and rich opportunities to explore mathematics in depth, and that is a very worthwhile thing for them to do.

Connect Mathematics to the World Using Mathematical Modeling

One main reason school children give for not liking mathematics is its abstract nature and per-ceived irrelevance to the world. This is a sad reflection on the math teaching they receive in schools, because math is everywhere and all around us. In fact, it is so critical to successful func-tioning in life that it has been termed the new "civil right"—essential for people to function well in society (Moses & Cobb, 2001). When I interviewed a group of young people, all about 24 years old, who had received traditional math teaching in school, and I asked them about math in

Painted Cube

Imagine a $5 \times 5 \times 5$ cube that had been painted blue on the outside, with cubes made up of smaller $1 \times 1 \times 1$ cubes. Consider the questions:

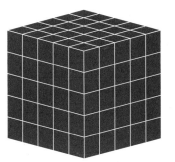

How many small cubes will have 3 blue faces?

How many small cubes will have 2 blue faces?

How many small cubes will have 1 blue face?

How many small cubes will have no paint on them?

Exhibit 9.7

FIGURE 9.8 A painted cube

A pyramid of cubes   A pyramid of pyramids   The 4th dimension

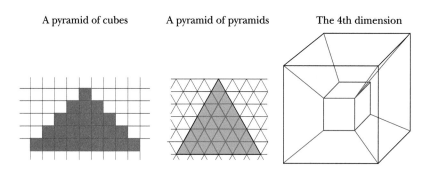

FIGURE 9.9   Extended cube problem

their lives and work, they expressed dismay at the math education they had received. The young adults said that they could see math all around them in the world now and they used it every day in their jobs, yet their school experiences of math had not given them any sense of the real nature of math and its importance to their future. They said that if they had only known that math wasn't a dead and irrelevant subject—that it would actually be essential to their adult lives—it would have made a big difference to their motivation in math classes in school.

The need to make mathematics interesting and connected to the real world has often resulted in publishers putting mathematics into what I call "pseudo contexts" (Boaler, 2015a) intended to represent reality. Students work on fake real-world problems that are far from reality, such as trains speeding toward each other on the same track. These contexts do not help students know that math is a useful subject. They give the opposite impression to students, as they show mathematics to be other-worldly and unreal. To be successful in a fake real-world problem, students are asked to engage as though the questions are real, at the same time as ignoring everything they know about the real-world situation. For example, consider these typical questions:

> Joe can do a job in 6 hours and Charlie can do the same job in 5 hours. What part of the job can they finish by working together for 2 hours?

> A restaurant charges $2.50 for 1/8 of a quiche. How much does a whole quiche cost?

> A pizza is divided into fifths for 5 friends at a party. Three of the friends eat their slices, but then 4 more friends arrive. What fractions should the remaining 2 slices be divided into? (Boaler, 2015a)

These questions all come from published textbooks and are typical of the questions students work with in math class. But they are all nonsensical. Everyone knows that people work at a different rate together than when alone, restaurants charge a different proportional price for food that is sold in bulk, and if extra friends arrive at a party more pizza is ordered—the remaining slices are not subdivided into small fractions. The cumulative effect of students working with pseudo contexts is that they come to think of math as irrelevant. In fact, for many students, they know that when they walk into math class they are walking into *Mathland,* a strange and mysterious place that requires them to leave their common sense at the door.

How then do we help students see the widespread use and applicability of mathematics without using pseudo contexts? The world is full of fascinating examples of situations that we can make sense of with mathematics. My online class helped students to see this by showing them the mathematics in snowflakes, in the work of spiders, in juggling and dancing, and in the calls of dolphins. The mathematics spanned from elementary to high levels of high school (Stanford Online Lagunita, 2014). Not all mathematics problems can or should be placed into a real-world context, as some of the greatest problems that help students learn important quantitative thinking are without a context. But it is important to have students see the applicability of mathematics and work with real-world variables for at least some of the time.

Conrad Wolfram urges viewers of his TED talk to view mathematics as a subject that centers around the posing of questions and forming of mathematical models (Wolfram, 2010). He highlights the act of modeling as central to the mathematics of the world. The Common Core Standards also highlight modeling, a mathematical practice standard.

MP 4: Model with Mathematics

One of the most important contributions of the Common Core State Standards (CCSS), in my view, is their inclusion of mathematical practices—the actions that are important to mathematics, in which students need to engage as they learn mathematics knowledge. "Modeling with Mathematics" is one of the 8 Mathematics Practices Standards (see box).

CCSS.MATH.PRACTICE.MP4 MODEL WITH MATHEMATICS.

Mathematically proficient students can apply the mathematics they know to solve problems arising in everyday life, society, and the workplace. In early grades, this might be as simple as writing an addition equation to describe a situation. In middle grades, a student might apply proportional reasoning to plan a school event or analyze a problem in the community. By high school, a student might use geometry to solve a design problem or use a function to describe how one quantity of interest depends on another. Mathematically proficient students who can apply what they know are comfortable making

(continued)

(continued)

assumptions and approximations to simplify a complicated situation, realizing that these may need revision later. They are able to identify important quantities in a practical situation and map their relationships using such tools as diagrams, two-way tables, graphs, flowcharts and formulas. They can analyze those relationships mathematically to draw conclusions. They routinely interpret their mathematical results in the context of the situation and reflect on whether the results make sense, possibly improving the model if it has not served its purpose.

Source: Common Core State Standards Initiative, 2015.

The act of modeling can be thought of as the simplification of any real-world problem into a pure mathematical form that can help to solve the problem. Modeling happens all through mathematics, but students have not typically been aware that they are modeling or asked to think about the process.

Ron Fedkiw is an applied mathematician at Stanford who specializes in computer-generated special effects. His mathematical models have created the special effects in award-winning movies such as *Pirates of the Caribbean: Dead Man's Chest* and *Star Wars: Episode III—Revenge of the Sith*. Fedkiw trained in pure mathematics until he was 23 and then moved to applied mathematics. As part of his work, he designs new algorithms that rotate objects, mimic collisions, and "mathematically stitch together slices of a falling water drop."

Mathematical modeling is also used in criminal cases and has helped to solve high-profile murder cases. *NUMB3RS* is a successful television show that features an FBI agent who often gets help from his mathematician brother. The first episode of *NUMB3RS* featured the true story of a brutal serial killer. FBI agents had been keeping track of murder locations on a map but could not see any patterns. The FBI agent in the show was stumped but recalled that his mathematician brother talked all the time about mathematics being the study of patterns. He asked his brother for help. The mathematician worked by inputting key information about serial killers, such as the fact that they tend to strike close to their home but not too close, and they leave a buffer zone within which they won't strike. He discovered that he could capture the pattern of the crosses with a simplified mathematical model. The model showed a "hot zone" indicating the areas in which the killer was likely to live. The FBI agents set to work investigating men of a certain age who lived in the zone, and the case was eventually solved. This episode was based on the work of the real-life mathematician Kim Rossmo, who developed a process of criminal geographic targeting (CGT) using mathematical models, a process used by police departments around the world.

When we ask students to take a problem from the world, based on real data and constraints, and solve it using mathematics, we are asking them to model the situation. As Wolfram says, students should encounter or find a real-world problem, set up a model to solve it, run some calculations (the part that can be done by a calculator or computer), and then see whether their answer solves the problem or the model needs to be refined. He points out that students currently spend 80% of the time they spend in math classrooms performing calculations, when they should instead be working on the other three parts of mathematics—setting up models, refining them, and using them to solve real problems.

In algebra classes students are often asked to compute rather than set up a model using algebra. For example, consider this problem:

> A man is on a diet and goes into a shop to buy some turkey slices. He is given 3 slices which together weigh 1/3 of a pound but his diet says that he is allowed to eat only 1/4 of a pound. How much of the 3 slices he bought can he eat while staying true to his diet?

This is a difficult problem for many people. But the difficulties most people face are not in the calculations; they arise in the setting up of a model to solve the problem. I have written elsewhere about the elegant visual solutions young children produce to solve this problem (Boaler, 2015a); this is the solution one fourth grader produced:

> If 3 slices is 1/3 of a pound then a pound is 9 slices (see Figure 9.10)

> If he can have one quarter of a pound, he can have one fourth of that (see Figure 9.11) … which is 2 1/4 slices.

By contrast, adults struggled to answer the problem, either incorrectly multiplying $\frac{1}{3} \times \frac{1}{4}$ or trying to use algebra but not remembering how. To use algebra, they needed to say

$$3 \text{ slices} = \tfrac{1}{3} \text{ pound}$$

$$x \text{ slices} = \tfrac{1}{4} \text{ pound}$$

and then cross multiply, so that $\frac{1}{3}x = \frac{3}{4}$ and $x = \frac{9}{4}$.

The adults who were given this problem experienced difficulty setting up the model and creating an expression. Despite years of work in algebra classes, students get very little experience with interpreting situations and setting up models. Students are trained to move variables around on a page and solve many expressions, but rarely do they set up problems. This is the important process Wolfram talks about—the setting up of a model.

Students of all ages can engage in modeling. For example, students in kindergarten class can be asked to make a seating map for all the children in the class, so that they can all fit on the carpet. They can represent each child by a shape or object and find a good way for all the children

FIGURE 9.10   Nine slices

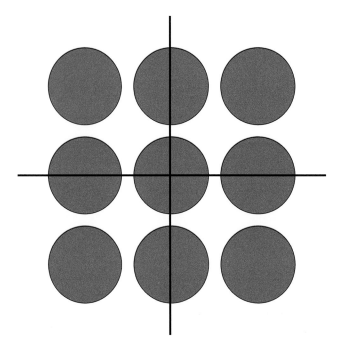

FIGURE 9.11   Nine slices divided into quadrants

to sit on the carpet. This is an example of modeling a situation, in this instance with shapes or objects representing more complex beings (the young children!) (Youcubed at Stanford University, 2015b; www.youcubed.org/task/moving-colors/).

A mathematical model often offers more simplicity than the real situation. In the kindergarten example, the shapes representing children do not take account of their size or movement. In the turkey slice example, the slices are assumed to be of equal size and weight with no variation.

A nice modeling question that students can work on in middle or high school is the famous tethered goat problem. The extended version of the problem in Exhibit 9.8 was written by Cathy Williams.

This question is set in a context that isn't real, but it is a context that invites students to consider aspects of the real situation and use them in their thinking. Students will probably wonder

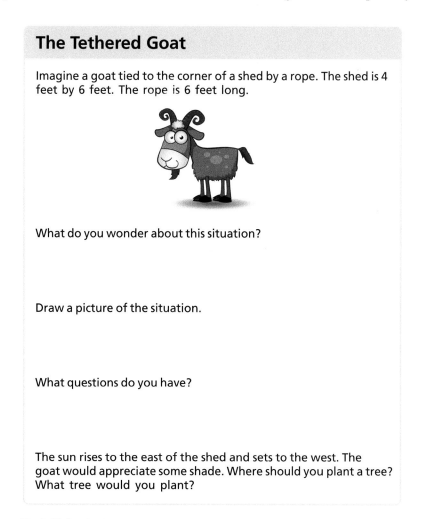

The Tethered Goat

Imagine a goat tied to the corner of a shed by a rope. The shed is 4 feet by 6 feet. The rope is 6 feet long.

What do you wonder about this situation?

Draw a picture of the situation.

What questions do you have?

The sun rises to the east of the shed and sets to the west. The goat would appreciate some shade. Where should you plant a tree? What tree would you plant?

Exhibit 9.8

about the space the goat has to move around in. They or the teacher could suggest adding some fencing. A nice extension is to ask students to decide how they would arrange 60 1-foot fences to maximize the additional area, which is a lovely, rich problem that I described in Chapter 5. When students think about planting the tree they might wonder what would happen if the goat ate the tree? What would be the best tree to plant? Where would you plant the tree so the goat could not eat it but would benefit from the shade?

This is a mathematical situation that has plenty of room for students to ask rich questions and investigate them. They would need to model the situation and construct representations, two important mathematical practices (see Figure 9.12).

A nice way to use real data is to ask students to work with real numbers and data from magazines, newspapers, and the Internet. For example, an activity I like that also teaches students about issues of social justice is one that asks them to form groups in the class to represent the different continents in the world. The groups then investigate how many cookies their group would get if the cookies represent the proportion of the world's wealth in their continent (see Exhibit 9.9). Students will model, reason, and apply knowledge as well as learn real and important information about the world and the way wealth is distributed, which will be especially real to them if it is translated into cookies they can eat. As students in some parts of the world get very few cookies, it is best to bring along spares to even out the cookie eating afterwards!

Olympic and other sporting data offer a wealth of opportunities for mathematical questioning and thinking. It is important when drawing from sporting data to be conscious of gender equity. Exhibit 9.10 presents a question I like, that again involves setting up a mathematical model:

My advice in bringing the real world into the classroom is to use real data and situations and to give a context only when it is helpful. Be sure it does not involve students suspending their sense making and stepping into *Mathland*.

The PISA team at the Organisation for Economic Co-operation and Development (OECD) conducted an interesting and useful analysis of the strengths and weaknesses U.S. students showed in the PISA international mathematics assessments. They found that the U.S. students' weaknesses were related to the artificial contexts used in classrooms, which do not teach students

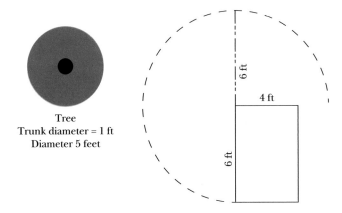

Tree
Trunk diameter = 1 ft
Diameter 5 feet

6 ft

4 ft

6 ft

FIGURE 9.12   Tethered goat model

World Wealth Simulation

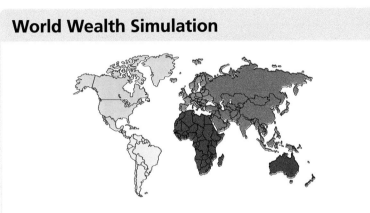

1. Find the percentage of the world's population living on each continent.

2. Calculate the number of people in our class who would correspond to the percentages found.

3. Calculate the percentage of the world's wealth for each continent.

4. Calculate the wealth of each continent in cookies.

TABLE 1 World Wealth Data

Continent	Population (in millions) 2000	Percent of Population	Wealth (GDP in trillions of dollars)	Percentage of Wealth
Africa	1,136		2.6	
Asia	4,351		18.5	
N. America	353		20.3	
S. America	410		4.2	
Europe	741		24.4	
Oceania/Aust.	39		1.8	
Total	7,030	100%	71.8	100%

Sources: Population data according to Population Reference Bureau (prb.org). Wealth data according to International Monetary Fund.

Exhibit 9.9

to use real-world variables but instead teach them to ignore them. Their recommendations for ways to engage students to encourage success are helpful:

It seems that the U.S. students have particular strengths in cognitively less-demanding mathematical skills and abilities, such as extracting single values from diagrams or handling well-structured formulae. And they have particular weaknesses in demanding skills and abilities, such as taking real world situations

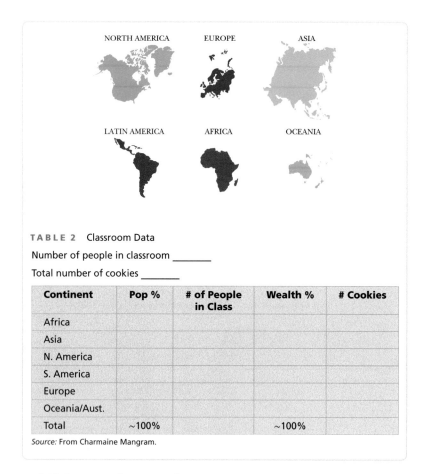

TABLE 2   Classroom Data

Number of people in classroom _____

Total number of cookies _____

Continent	Pop %	# of People in Class	Wealth %	# Cookies
Africa				
Asia				
N. America				
S. America				
Europe				
Oceania/Aust.				
Total	~100%		~100%	

Source: From Charmaine Mangram.

Exhibit 9.9   (Continued)

seriously, transferring them into mathematical terms and interpreting mathematical aspects in real world problems. These are tasks where the well-known superficial classroom strategy "Don't care about the context, just extract the numbers from the text and do some obvious operations" is bound to fail. This strategy is popular all over the world and frequently helps pupils and students to survive in school mathematics and to pass examinations. However, in a typical PISA mathematical literacy task, the students have to use the mathematics they have learned in a well-founded manner. The American students obviously have particular problems with such tasks. (…) When it comes to the implications of these findings, one clear recommendation would be to focus much more on higher-order activities such as those involved in mathematical modeling (understanding real world situations, transferring them into mathematical models, and interpreting mathematical results), without neglecting the basic skills needed for these activities. (Organisation for Economic Co-operation and Development [OECD], 2013)

Soccer Goalie

If you are a soccer goalie, and an attacking player from the other team has broken away from the others and is running toward you, where is the best place for you to stand? Try mapping out different positions depending on the location of the attacking player when she shoots.

Exhibit 9.10

The PISA team observed a phenomenon that comes from the weakness of the questions given to students in the United States. They noted that students' tendency to ignore contexts and just use numbers resulted in failure on their questions. This reflects the low quality of the questions used in textbooks across the United States, with their fake contexts. Sadly, the strategies U.S. students typically learn from math class will be similarly unhelpful when they enter the world of work. Students need to be engaged in math class with questions that require them to consider a real situation, use real-world variables, and engage with data from the world. They need to learn to set up mathematical models from the situations and to problem solve, a process that is both engaging and extremely important for their future.

Encourage Students to Pose Questions, Reason, Justify, and Be Skeptical

FIGURE 9.13 Bracelets for sale

The first thing a mathematician has to do is pose an interesting question. This mathematical practice is virtually absent in math classrooms, yet it is central to mathematical work. Nick Foote is a wonderful third-grade public school teacher and friend who taught both of my daughters, which has given us the opportunity to have many discussions about math together. In Nick's class he sometimes gives situations and invites students to come up with their own mathematical questions. I was visiting Nick's class one day when he gave this situation (see Figure 9.13).

You want to buy some Wonder Loom bracelets. You go to the Rainbow Zen Garden Store and find these options.

Two-color bracelets—$0.50 each or 3 for $1.00

Multicolor bracelets—$1.00 each or 3 for $2.50

Supplies to create your own bracelets:

600-count bag of rubber bands—$3.00

or 4 bags—$10

600-count bag of glow-in-the-dark rubber bands—$5.00

Wonder loom—$5.00

He then asked groups of students to discuss the situation and pose questions. Exhibit 9.11 is a handout Nick often uses.

The students excitedly set about wondering about questions such as: why are the bracelets so expensive to buy? They were helped in working this out by finding out how much it would cost to make a bracelet from the materials, and then thinking about the cost of selling through a shop. These were real questions from the students, eliciting higher engagement and learning.

When students move into employment, in our high-tech world, one very important job they will be asked to do is to pose questions of situations and of large data sets. Increasingly, companies are dealing with giant data sets, and the people who can ask creative and interesting questions of the data will be highly valued in the workplace. In my own teaching experience, when I have asked students in classrooms to consider a situation and pose their own question, they have become instantly engaged, excited to draw on their own thinking and ideas. This is an idea for math classrooms that is very easy to implement and needs to be used only some of the time. Students should

We Wonder

Team Members:

Date:

We wonder

_____

_____

Use pictures, numbers, and words to show how you answered your question.

We want to investigate

_____

_____

Use pictures, numbers, and words to show how you answered your question.

Source: From Nick Foote.

Exhibit 9.11

be able to experience this in school so that they are prepared to use it later in their mathematical lives.

When Conrad Wolfram discusses his role as an employer, he says that he does not need people who can calculate fast, as computers do that work. He needs people who can make conjectures and talk about their mathematical pathways. It is so important that employees describe their mathematical pathways to others, in teams, because others can then use those pathways in their own work and investigations and can also see if there are errors in thinking or logic. This is the core of mathematical work; it is called reasoning.

I speak to many groups of parents about Common Core math, and I am often asked, especially by the parents of high-achieving students, "Why should my child discuss his work in a group, when he can get the answers quickly on his own?" I explain to parents that explaining one's work is a mathematical practice, called reasoning, that is at the heart of the discipline. When students offer reasons for their mathematical ideas and justify their thinking, they are engaging in mathematics. Scientists work by proposing theories and looking for cases that prove or disprove their theory. Mathematicians propose theories and reason about their mathematical pathways, justifying the logical connections they have made between ideas (Boaler, 2013c).

Chapter Five introduced a classroom strategy of inviting students to be skeptics, which prompts students to push each other to high levels of reasoning. This is an excellent way to teach students reasoning and for them to take on the role of skeptic, which students enjoy. As I described in Chapter Five, reasoning is not only a central mathematical practice but also a classroom practice that promotes equity, as it helps all students get access to ideas. When students act as a skeptic, they get an opportunity to question other students without having to take on the role of someone who doesn't understand.

Teach with Cool Technology and Manipulatives

As we invite students to enter a world in which mathematics is open, visual, and creative, many forms of technology and manipulatives are helpful. Cuisenaire rods, multilink cubes, and pattern blocks are helpful for students at all levels of mathematics; I use them in my undergraduate teaching at Stanford. Chapter Four reviewed a range of apps and games that also invite students into visual and conceptual thinking. I focused in that chapter on number, but many good apps also enable students to explore geometric ideas in two and three dimensions, allowing students to move angles and lines in order to explore relationships. This is important and powerful thinking that cannot be done with pen and paper. Geometry Pad for iPad and GeoGebra both allow teachers and students to make their own dynamic demos, encouraging students, for example, to investigate geometric and algebraic ideas such as $y = mx + b$ and trig ratios, dynamically and visually. Geometry Pad is made by Bytes Arithmetic LLC, and the basic version is free.

Other apps, such as Tap Tap Blocks, help students build in three dimensions, making and solving spatial patterns and algebraic patterns (see Figure 9.14). Students can place and spin objects in a 3-D simulated space. Tap Tap Blocks is a free app made by Paul Hangas; it runs on iOS.

A nice activity to try with Tap Tap Blocks is to ask students to try and make a shape from screen shots of different views and then challenge a friend by making their own problem. For example:

> Can you build this shape that has 1 orange, 1 yellow, 1 dark blue, 2 green, 2 light blue, 2 red, and 3 purple blocks? Here are some different screen shots of the shape from different angles (See Figure 9.15).

These different apps offer productive ways to engage students with conceptual and visual thinking, but they are not the only apps, games, or sites that do this. There are millions of math apps and games that claim to help students, but few that build on research on learning, showing mathematics as a conceptual and visual subject. My advice is to be discerning when you choose technology to engage your students, using those that motivate students to think and make connections, not to work at speed on procedures and calculations.

Mathematics is a broad, multidimensional subject, and when teachers embrace the multidimensionality of mathematics, in their teaching and through their assessment, many more

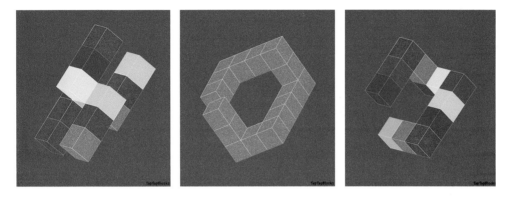

FIGURE 9.14 Tap Tap Blocks

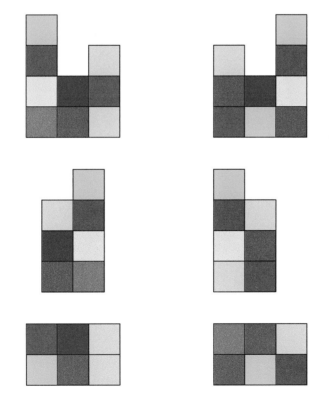

FIGURE 9.15 A Tap Tap Blocks shape from six different angles

students can gain access to mathematics and be excited by it. When we open mathematics, we broaden the number and range of students who can engage and do well. This is not an artificial broadening or dumbing-down of mathematics; rather, it is a broadening that brings school mathematics closer to real mathematics and the mathematics of the world.

Conclusion

Teachers, parents and leaders have the opportunity to set students on a growth mindset mathematics pathway that will bring them greater accomplishment, happiness, and feelings of self-worth throughout their lives. We need to free our young people from the crippling idea that they must not fail, that they cannot mess up, that only some students can be good at math, and that success should be easy and not involve effort. We need to introduce them to creative, beautiful mathematics that allows them to ask questions that have not been asked, and to think of ideas that go beyond traditional and imaginary boundaries. We need our students to develop *Growth Mathematical Mindsets*. I hope that this book has given you some ideas that can start or reinvigorate your own journey into creative and growth mathematics and mindset that will continue throughout your life. When we encourage open mathematics and the learning messages that support it, we develop our own intellectual freedom, as teachers and parents, and inspire that freedom in others.

Thank you for setting out on this journey with me. Now it is time for you to invite others onto the pathways you have learned, inviting them to be the people they should be, free from artificial rules, and inspired by the knowledge they have unlimited mathematics potential. For we can all open mathematics, and give students the chance to ask their own questions and bring their own natural creativity and curiosity to the foreground as they learn. If we give students this rich, creative, growth mathematics experience, then we change them as people and the ways they interact with the world.

When we set students free, beautiful mathematics follows.

REFERENCES

Abiola, O., & Dhindsa, H. S. (2011). Improving classroom practices using our knowledge of how the brain works. *International Journal of Environmental & Science Education, 7*(1), 71–81.

Baker, D. P., & LeTendre, G. K. (2005). *National differences, global similarities: World culture and the future of schooling.* Stanford, CA: Stanford University Press.

Ball, D. L. (1993). With an eye on the mathematical horizon: Dilemmas of teaching elementary mathematics. *The Elementary School Journal, 93*(4), 373–397.

Beaton, A. E., & O'Dwyer, L. M. (2002). Separating school, classroom and student variances and their relationship to socio-economic status. In D. F. Robitaille & A. E. Beaton (Eds.), *Secondary analysis of the TIMSS data.* Dordrecht: Kluwer Academic Publishers.

Beilock, L. S., Gunderson, E. A., Ramirez, G., & Levine, S. C. (2009). Female teachers' math anxiety affects girls' math achievement. *Proceedings of the National Academy of Sciences, 107*(5), 1860–1863.

Beilock, S. (2011). *Choke: What the secrets of the brain reveal about getting it right when you have to.* New York: Free Press.

Black, P., Harrison, C., Lee, C., Marshall, B., & Wiliam, D. (2002). *Working inside the black box: Assessment for learning in the classroom.* London: Department of Education & Professional Studies, King's College.

Black, P. J., & Wiliam, D. (1998a, October). Inside the black box: Raising standards through classroom assessment. *Phi Delta Kappan,* 139–148.

Black, P. J., & Wiliam, D. (1998b). Assessment and classroom learning. *Assessment in Education, 5*(1), 7–74.

Blackwell, L., Trzesniewski, K., & Dweck, C. S. (2007). Implicit theories of intelligence predict achievement across an adolescent transition: A longitudinal study and an intervention. *Child Development, 78*(1), 246–263.

Boaler, J. (1997). When even the winners are losers: Evaluating the experiences of "top set" students. *Journal of Curriculum Studies, 29*(2), 165–182.

Boaler, J. (1998). Open and closed mathematics: Student experiences and understandings. *Journal for Research in Mathematics Education, 29*(1), 41–62.

Boaler, J. (2002a). *Experiencing school mathematics: Traditional and reform approaches to teaching and their impact on student learning* (revised, expanded edition). Mahwah, NJ: Erlbaum.

Boaler, J. (2002b). Paying the price for "sugar and spice": Shifting the analytical lens in equity research. *Mathematical Thinking and Learning, 4*(2&3), 127–144.

Boaler, J. (2005). *The "psychological prisons" from which they never escaped: The role of ability grouping in reproducing social class inequalities.* Paper presented at the FORUM.

Boaler, J. (2008). Promoting "relational equity" and high mathematics achievement through an innovative mixed ability approach. *British Educational Research Journal, 34*(2), 167–194.

Boaler, J. (2013a). Ability and mathematics: The mindset revolution that is reshaping education. *FORUM, 55*(1), 143–152.

Boaler, J. (2013b). Ability grouping in mathematics classrooms. In S. Lerman (Ed.), *International encyclopedia of mathematics education*: New York: Springer.

Boaler, J. (2013c, November 12). The stereotypes that distort how Americans teach and learn math. *Atlantic.*

Boaler, J. (2014a, April 28). *Changing the conversation about girls and STEM*. Washington, DC: The White House. Retrieved from http://www.youcubed.org/wp-content/uploads/Youcubed-STEM-white-house.pdf

Boaler, J. (2014b). The mathematics of hope—Moving from performance to learning in mathematics classrooms. Retrieved from https://www.youcubed.org/the-mathematics-of-hope-moving-from-performance-to-learning-in-mathematics-classrooms/

Boaler, J. (2014c). Fluency without fear: Research evidence on the best ways to learn math facts. YouCubed at Stanford University. Retrieved from http://www.youcubed.org/wp-content/uploads/2015/03/FluencyWithoutFear-2015.pdf

Boaler, J. (2015a). *What's math got to do with it? How teachers and parents can transform mathematics learning and inspire success*. New York: Penguin.

Boaler, J. (2015b, May 7). Memorizers are the lowest achievers and other Common Core math surprises. *The Hechinger Report*. Retrieved from http://hechingerreport.org/memorizers-are-the-lowest-achievers-and-other-common-core-math-surprises/

Boaler, Jo. (2014). Ability grouping in mathematics classrooms. In S. Lerman (Ed.), *Encyclopedia of mathematics education* (pp. 1–5). Dordrecht: Springer Science+Business Media. doi:10.1007/978-94-007-4978-8

Boaler, J., & Greeno, J. (2000). Identity, agency and knowing in mathematics worlds. In J. Boaler (Ed.), *Multiple perspectives on mathematics teaching and learning* (pp. 171–200). Westport, CT: Ablex Publishing.

Boaler, J., & Humphreys, C. (2005). *Connecting mathematical ideas: Middle school video cases to support teaching and learning*. Portsmouth, NH: Heinemann.

Boaler, J., Meyer, D., Selling, S. K., & Sun, K. (n.d.). The *Simpsons* sunblocker: Similarity and congruence through modeling, exploration, and reasoning. Youcubed at Stanford University. Retrieved from http://www.youcubed.org/wp-content/uploads/The-Sunblocker1.pdf

Boaler, J., & Sengupta-Irving, T. (2015). The many colors of algebra: Engaging disaffected students through collaboration and agency. *Journal of Mathematical Behavior*.

Boaler, J., & Staples, M. (2005). Transforming students' lives through an equitable mathematics approach: The case of Railside School. *Teachers College Record, 110*(3), 608–645.

Boaler, J., & Wiliam, D. (2001). "We've still got to learn!" Students' perspectives on ability grouping and mathematics achievement. In P. Gates (Ed.), *Issues in mathematics teaching*. London: RoutledgeFalmer.

Boaler, J., Wiliam, D., & Brown, M. (2001). Students' experiences of ability grouping—disaffection, polarisation and the construction of failure. *British Educational Research Journal, 26*(5), 631–648.

Bransford, J., Brown, A., & Cocking, R. (1999). *How people learn: Brain, mind, experience and school*. Washington, DC: National Academy Press.

Brousseau, G. (1984). The crucial role of the didactical contract in the analysis and construction of situations in teaching and learning mathematics. In H. G. Steiner (Ed.), *Theory of mathematics education* (pp. 110–119). Bielefeld Germany: Institut für Didactik der Mathematik der Universität Bielefeld.

Brousseau, G. (1997). *Theory of didactical situations in mathematics: Didactique des mathématiques* (1970–1990). New York, NY: Springer.

Bryant, A. (2013, June 19). In head-hunting, big data may not be such a big deal. *New York Times*. Retrieved from http://www.nytimes.com/2013/06/20/business/in-head-hunting-big-data-may-not-be-such-a-big-deal.html

Burris, C., Heubert, J., & Levin, H. (2006). Accelerating mathematics achievement using heterogeneous grouping. *American Educational Research Journal, 43*(1), 103–134.

Burton, L. (1999). The practices of mathematicians: What do they tell us about coming to know mathematics? *Educational Studies in Mathematics, 37*, 121–143.

Butler, R. (1987). Task-involving and ego-involving properties of evaluation: Effects of different feedback conditions on motivational perceptions, interest and performance. *Journal of Educational Psychology, 79*, 474–482.

Butler, R. (1988). Enhancing and undermining intrinsic motivation: The effects of task-involving and ego-involving evaluation on interest and performance. *British Journal of Educational Psychology, 58*, 1–14.

Challenge Success. (2012). Changing the conversation about homework from quantity and achievement to quality and engagement. Stanford, CA: Challenge Success. Retrieved from http://www.challengesuccess.org/wp-content/uploads/2015/07/ChallengeSuccess-Homework-WhitePaper.pdf

Cohen, E. (1994). *Designing groupwork*. New York: Teachers College Press.

Cohen, E., & Lotan, R. (2014). *Designing groupwork: Strategies for the heterogeneous classroom* (3rd ed.). New York: Teachers College Press.

Cohen, G. L., & Garcia, J. (2014). Educational theory, practice, and policy and the wisdom of social psychology. *Policy Insights from the Behavioral and Brain Sciences, 1*(1), 13–20.

Common Core State Standards Initiative. (2015). Standards for mathematical practice. Common Core State Standards Initiative. Retrieved from http://www.corestandards.org/Math/Practice/

Conner, J., Pope, D., & Galloway, M. K. (2009). Success with less stress. *Educational Leadership, 67*(4), 54–58.

Darling-Hammond, L. (2000). Teacher quality and student achievement. *Education Policy Analysis Archives, 8*, 1.

Deevers, M. (2006). *Linking classroom assessment practices with student motivation in mathematics*. Paper presented at the American Educational Research Association, San Francisco.

Delazer, M., Ischebeck, A., Domahs, F., Zamarian, L., Koppelstaetter, F., Siedentopf, C. M., … Felber, S. (2005). Learning by strategies and learning by drill—evidence from an fMRI study. *NeuroImage*, 839–849.

Devlin, K. (1997). *Mathematics: The science of patterns: The search for order in life, mind and the universe*: Scientific American Library: New York.

Devlin, K. (2001). *The math gene: How mathematical thinking evolved and why numbers are like gossip*: Basic Books: New York. (Originally published in 1997)

Devlin, K. (2006). *The math instinct: Why you're a mathematical genius (along with lobsters, birds, cats, and dogs)*: Basic Books: New York.

Dixon, A. (2002). Editorial. *FORUM, 44*(1), 1.

Duckworth, A., & Quinn, P. (2009). Development and validation of the short grit scale. *Journal of Personality Assessment, 91*(2), 166–174.

Duckworth, E. (1991). Twenty-four, forty-two and I love you: Keeping it complex. *Harvard Educational Review, 61*(1), 1–24.

Dweck, C. S. (2006a). Is math a gift? Beliefs that put females at risk. In W.W.S.J. Ceci (Ed.), *Why aren't more women in science? Top researchers debate the evidence*. Washington, DC: American Psychological Association.

Dweck, C. S. (2006b). *Mindset: The new psychology of success*. New York: Ballantine Books.

Eccles, J., & Jacobs, J. (1986). Social forces shape math attitudes and performance. *Signs, 11*(2), 367–380.

Elawar, M. C., & Corno, L. (1985). A factorial experiment in teachers' written feedback on student homework: Changing teacher behavior a little rather than a lot. *Journal of Educational Psychology, 77*(2), 162–173.

Elmore, R., & Fuhrman, S. (1995). Opportunity-to-learn standards and the state role in education. *Teachers College Record, 96*(3), 432–457.

Engle, R. A., Langer-Osuna, J., & McKinney de Royston, M. (2014). Towards a model of influence in persuasive discussions: Negotiating quality, authority, and access within a student-led argument. *Journal of the Learning Sciences, 23*(2), 245–268.

Esmonde, I., & Langer-Osuna, J. (2013). Power in numbers: Student participation in mathematical discussions in heterogeneous spaces. *Journal for Research in Mathematics Education, 44*(1), 288–315.

Feikes, D., & Schwingendorf, K. (2008). The importance of compression in children's learning of mathematics and teacher's learning to teach mathematics. *Mediterranean Journal for Research in Mathematics Education, 7*(2).

Flannery, S. (2002). *In code: A mathematical journey*: Workman Publishing Company: New York.

Fong, A. B., Jaquet, K., & Finkelstein, N. (2014). Who repeats Algebra I, and how does initial performance relate to improvement when the course is repeated? (REL 2015–059). Washington, DC: U.S. Department of Education, Institute of Education Sciences, National Center for Education Evaluation and Regional Assistance, Regional Educational Laboratory West. Retrieved from http://files.eric.ed.gov/fulltext/ED548534.pdf

Frazier, L. (2015, February 25). To raise student achievement, North Clackamas schools add lessons in perseverance. *Oregonian*/OregonLive. Retrieved from http://www.oregonlive.com/education/index.ssf/2015/02/to_raise_student_achievement_n.html

Galloway, M. K., & Pope, D. (2007). Hazardous homework? The relationship between homework, goal orientation, and well-being in adolescence. *Encounter: Education for Meaning and Social Justice, 20*(4), 25–31.

Girl Scouts of the USA with the National Center for Women & Information Technology (NCWIT). (2008). Evaluating promising practices in informal science, technology, engineering and mathematics (STEM) education for girls. Retrieved from http://www.ncwit.org/sites/default/files/legacy/pdf/NCWIT-GSUSAPhaseIIIReport_FINAL.pdf

Gladwell, M. (2011). *Outliers: The story of success*: Back Bay Books.

Good, C., Rattan, A., & Dweck, C. S. (2012). Why do women opt out? Sense of belonging and women's representation in mathematics. *Journal of Personality and Social Psychology, 102*(4), 700–717.

Gray, E., & Tall, D. (1994). Duality, ambiguity, and flexibility: A "proceptual" view of simple arithmetic. *Journal for Research in Mathematics Education, 25*(2), 116–140.

Gunderson, E. A., Gripshover, S. J., Romero, C., Dweck, C. S., Goldin-Meadow, S., & Levine, S. C. (2013). Parent praise to 1–3 year-olds predicts children's motivational frameworks 5 years later. *Child Development, 84*(5), 1526–1541.

Gutstein, E., Lipman, P., Hernandez, P., & de los Reyes, R. (1997). Culturally relevant mathematics teaching in a Mexican American context. *Journal for Research in Mathematic Education, 28*(6), 709–737.

Haack, D. (2011, January 31). Disequilibrium (I): Real learning is disruptive. A Glass Darkly (blog). Retrieved from http://blog4critique.blogspot.com/2011/01/disequilibrium-i-real-learning-is.html

Hersh, R. (1999). *What is mathematics, really?* Oxford, UK: Oxford University Press.

Horn, I. S. (2005). Learning on the job: A situated account of teacher learning in high school mathematics departments. *Cognition and Instruction, 23*(2), 207–236.

Humphreys, C., & Parker, R. (2015). *Making number talks matter: Developing mathematical practices and deepening understanding, grades 4–10.* Portland, ME: Stenhouse Publishers.

Jacob, W. (2015). [Math Acceleration]. Personal communication.

Jones, M. G., Howe, A., & Rua, M. J. (2000). Gender differences in students' experiences, interests, and attitudes toward science and scientists. *Science Education, 84*, 180–192.

Karni, A., Meyer, G., Rey-Hipolito, C., Jezzard, P., Adams, M., Turner, R., & Ungerleider, L. (1998). The acquisition of skilled motor performance: Fast and slow experience-driven changes in primary motor cortex. *PNAS, 95*(3), 861–868.

Khan, S. (2012). *The one world schoolhouse: Education reimagined.* New York: Twelve.

Kitsantas, A., Cheema, J., & Ware, W. H. (2011). Mathematics achievement: The role of homework and self-efficacy beliefs. *Journal of Advanced Academics, 22*(2), 310–339.

Klarreich, E. (2014, August 13). Meet the first woman to win math's most prestigious prize. *Quanta Magazine.* Retrieved from http://www.wired.com/2014/08/maryam-mirzakhani-fields-medal/

Kohn, A. (1999). *Punished by rewards: The trouble with gold stars, incentive plans, A's, praise, and other bribes.* New York: Mariner Books.

Kohn, A. (2000). *The schools our children deserve: Moving beyond traditional classrooms and "tougher standards.* New York: Mariner Books.

Kohn, A. (2008, September). Teachers who have stopped assigning homework (blog). Retrieved from http://www.alfiekohn.org/blogs/teachers-stopped-assigning-homework

Kohn, A. (2011, November). The case against grades. Retrieved from http://www.alfiekohn.org/article/case-grades/

Koonlaba, A. E. (2015, February 24). 3 visual artists—and tricks—for integrating the arts into core subjects. *Education Week Teacher.* Retrieved from http://www.edweek.org/tm/articles/2015/02/24/3-visual-artists-and-tricks-for-integrating-the-arts.html

Lakatos, I. (1976). *Proofs and refutations.* Cambridge, UK: Cambridge University Press.

Langer-Osuna, J. (2011). How Brianna became bossy and Kofi came out smart: Understanding the differentially mediated identity and engagement of two group leaders in a project-based mathematics classroom. *Canadian Journal for Science, Mathematics, and Technology Education, 11*(3), 207–225.

Lawyers' Committee for Civil Rights of the San Francisco Bay Area. (2013, January). Held back: Addressing misplacement of 9th grade students in Bay Area school math classes. Retrieved from www.lccr.com

Lee, D. N. (2014, November 25). Black girls serving as their own role models in STEM. *Scientific American.* Retrieved from http://blogs.scientificamerican.com/urban-scientist/2014/11/25/black-girls-serving-as-their-own-role-models-in-stem/

Lee, J. (2002). Racial and ethnic achievement gap trends: Reversing the progress toward equity? *Educational Researcher, 31*(1), 3–12.

Lemos, M. S., & Veríssimo, L. (2014). The relationships between intrinsic motivation, extrinsic motivation, and achievement, along elementary school. *Procedia – Social and Behavioral Sciences, 112*, 930–938.

Leslie, S.-J., Cimpian, A., Meyer, M., & Freeland, E. (2015). Expectations of brilliance underlie gender distributions across academic disciplines. *Science, 347*(6219), 262–265.

Lupton, T., Pratt, S., & Richardson, K. (2014). Exploring long division through division quilts. *Centroid, 40*(1), 3–8.

Maguire, E., Woollett, K., & Spiers, H. (2006). London taxi drivers and bus drivers: A structural MRI and neuropsychological analysis. *Hippocampus, 16*(12), 1091–1101.

Mangels, J. A., Butterfield, B., Lamb, J., Good, C., & Dweck, C. S. (2006). Why do beliefs about intelligence influence learning success? A social cognitive neuroscience model. *Social Cognitive and Affective Neuroscience, 1*(2), 75–86.

Mathematics Teaching and Learning to Teach, University of Michigan. (2010). SeanNumbers-Ofala video [Online]. Retrieved from http://deepblue.lib.umich.edu/handle/2027.42/65013

McDermott, R. P. (1993). The acquisition of a child by a learning disability. In S. Chaiklin & J. Lave (Eds.), *Understanding practice: Perspectives on activity and context* (pp. 269–305). Cambridge, UK: Cambridge University Press.

McKnight, C., Crosswhite, F. J., Dossey, J. A., Kifer, J. O., Swafford, J. O., Travers, K. J., & Cooney, T. J. (1987). *The underachieving curriculum—Assessing US school mathematics from an international perspective*. Champaign, IL: Stipes Publishing.

Mikki, J. (2006). *Students' homework and TIMSS 2003 mathematics results*. Paper presented at the International Conference, Teaching Mathematics Retrospective and Perspective.

Moser, J., Schroder, H. S., Heeter, C., Moran, T. P., & Lee, Y. H. (2011). Mind your errors: Evidence for a neural mechanism linking growth mindset to adaptive post error adjustments. *Psychological Science, 22*, 1484–1489.

Moses, R., & Cobb, J. C. (2001). *Radical equations: Math, literacy and civil rights*. Boston: Beacon Press.

Mueller, C. M., & Dweck, C. S. (1998). Praise for intelligence can undermine children's motivation and performance. *Journal of Personality and Social Psychology, 75*(1), 33–52.

Murphy, M. C., Garcia, J. A., & Zirkel, S. (in prep). *The role of faculty mindsets in women's performance and participation in STEM*.

Nasir, N. S., Cabana, C., Shreve, B., Woodbury, E., & Louie, N. (Eds.). (2014). *Mathematics for equity: A framework for successful practice*. New York: Teachers College Press.

Noguchi, S. (2012, January 14). Palo Alto math teachers oppose requiring Algebra II to graduate. San Jose Mercury News. Retrieved from http://www.mercurynews.com/ci_19748978

Organisation for Economic Co-operation and Development (OECD). (2013). Lessons from PISA 2012 for the United States, strong performers and successful reformers in education. Paris: OECD.

Organisation for Economic Co-operation and Development (OECD). (2015). The ABC of gender equality in education: Aptitude, behaviour, confidence. A Program for International Student Assessment (PISA) Report. Paris: OECD Publishing.

Paek, P., & Foster, D. (2012). *Improved mathematical teaching practices and student learning using complex performance assessment tasks*. Paper presented at the National Council on Measurement in Education (NCME), Vancouver, BC, Canada.

Park, J., & Brannon, E. (2013). Training the approximate number system improves math proficiency. *Association for Psychological Science*, 1–7.

Parrish, S. (2014). *Number talks: Helping children build mental math and computation strategies, grades K–5, updated with Common Core Connections*. Sausalito, CA: Math Solutions.

Piaget, J. (1958). *The child's construction of reality*. London: Routledge & Kegan Paul.

Piaget, J. (1970). Piaget's theory. In P. H. Mussen (Ed.), *Carmichael's manual of child psychology*. New York: Wiley.

Picciotto, H. (1995). *Lab gear activities for Algebra I*. Mountain View, CA: Creative Publications.

Program for International Student Assessment (PISA). (2012). PISA 2012 results in focus. What 15-year-olds know and what they can do with what they know. Paris, France: OECD.

Program for International Student Assessment (PISA). (2015). Does homework perpetuate inequities in education? *PISA in Focus 46*. Retrieved from http://www.oecd-ilibrary.org/docserver/download/ 5jxrhqhtx2xt.pdf?expires=1426872704&id=id&accname=guest&checksum=D4D915DF09179 A5FB60344FEA43FE5E7

Pulfrey, C., Buchs, C., & Butera, F. (2011). Why grades engender performance-avoidance goals: The mediating role of autonomous motivation. *Journal of Educational Psychology, 103*(3), 683–700. Retrieved from http://www.researchgate.net/profile/Fabrizio_Butera/publication/232450947_ Why_grades_engender_performance-avoidance_goals_The_mediating_role_of_autonomous_ motivation/links/02bfe50ed4ebfd0670000000.pdf

Reeves, D. B. (2006). *The learning leader: How to focus school improvement for better results*. Alexandria, VA: Association for Supervision & Curriculum Development.

Romero, C. (2013). *Coping with challenges during middle school: The role of implicit theories of emotion* (Doctoral dissertation). Stanford University, Stanford, CA. Retrieved from http://purl.stanford .edu/ft278nx7911

Rose, H., & Betts, J. R. (2004). The effect of high school courses on earnings. *Review of Economics and Statistics, 86*(2), 497–513.

Schmidt, W. H., McKnight, C. C., Cogan, L. S., Jakwerth, P. M., Houang, R. T., Wiley, D. E., … Raizen, S. A. (2002). *Facing the consequences: Using TIMSS for a closer look at US mathematics and science education*. Dordrecht, Netherlands: Kluwer Academic Publishers.

Schmidt, W. H., McKnight, C. C., & Raizen, S. A. (1997). *A splintered vision: An investigation of US science and mathematics education*. Dordrecht, Netherlands: Kluwer Academic Publishers.

Schwartz, D., & Bransford, J. (1998). A time for telling. *Cognition and Instruction, 16*(4), 475–522.

Schwartz, L. (2001). *A mathematician grappling with his century*. Basel, Switzerland: Birkhäuser.

Seeley, C. (2009). *Faster isn't smarter: Messages about math, teaching, and learning in the 21st century*. Sausalito, CA: Math Solutions.

Seeley, C. (2014). *Smarter than we think: More messages about math, teaching, and learning in the 21st century*. Sausalito, CA: Math Solutions.

Selling, S. K. (2015). Learning to represent, representing to learn. *Journal of Mathematical Behavior*.

Silva, E., & White, T. (2013). *Pathways to improvement: Using psychological strategies to help college students master developmental math*. Stanford, CA: Carnegie Foundation for the Advancement of Teaching.

Silver, E. A. (1994). On mathematical problem posing. *For the Learning of Mathematics, 14*(1), 19–28.

Sims, P. (2011, August 6). Daring to stumble on the road to discovery. *New York Times*. Retrieved from http://www.nytimes.com/2011/08/07/jobs/07pre.html?_r=0

Solomon, Y. (2007). Not belonging? What makes a functional learner identity in undergraduate mathematics? *Studies in Higher Education, 32*(1), 79–96.

Stanford Center for Professional Development. (n.d.). How to learn math: For teachers and parents. http://scpd.stanford.edu/instanford/how-to-learn-math.jsp

Stanford Online Lagunita. (2014). How to learn math: For students. Stanford University. Retrieved from https://lagunita.stanford.edu/courses/Education/EDUC115-S/Spring2014/about

Steele, C. (2011). *Whistling Vivaldi: How stereotypes affect us and what we can do*. New York: Norton.

Stigler, J., & Hiebert, J. (1999). *The teaching gap: Best ideas from the world's teachers for improving education in the classroom*. New York: Free Press.

Stipek, D. J. (1993). *Motivation to learn: Integrating theory and practice*. New York: Pearson.

Supekar, K., Swigart, A. G., Tenison, C., Jolles, D. D., Rosenberg-Lee, M., Fuchs, L., & Menon, V. (2013). Neural predictors of individual differences in response to math tutoring in primary-grade school children. *Proceedings of the National Academy of Sciences, 110*(20), 8230–8235.

Thompson, G. (2014, June 2). Teaching the brain to learn. *THE Journal*. Retrieved from http://thejournal.com/articles/2014/06/02/teaching-the-brain-to-learn.aspx

Thurston, W. (1990). Mathematical education. *Notices of the American Mathematical Society, 37*(7), 844–850.

Tobias, S. (1978). *Overcoming math anxiety*. New York: Norton.

Treisman, U. (1992). Studying students studying calculus: A look at the lives of minority mathematics students in college. *College Mathematics Journal, 23*(5), 362–372.

Vélez, W. Y., Maxwell, J. W., & Rose, C. (2013). Report on the 2012–2013 new doctoral recipients. *Notices of the American Mathematical Society, 61*(8), 874–884.

Wang, J. (1998). Opportunity to learn: The impacts and policy implications. *Educational Evaluation and Policy Analysis, 20*(3), 137–156. doi:10.3102/01623737020003137

Wenger, E. (1998). *Communities of practice: Learning, meaning and identity*. Cambridge: Cambridge University Press.

White, B. Y., & Frederiksen, J. R. (1998). Inquiry, modeling, and metacognition: Making science accessible to all students. *Cognition and Instruction, 16*(1), 3–118.

Wolfram, C. (2010, July). Teaching kids real math with computers. TED Talks. Retrieved from http://www.ted.com/talks/conrad_wolfram_teaching_kids_real_math_with_computers?language=en

Woollett, K., & Maguire, E. A. (2011). Acquiring "The Knowledge" of London's layout drives structural brain changes. *Current Biology, 21*(24), 2109–2114.

Youcubed at Stanford University. (2015a). Making group work equal. Stanford Graduate School of Education. Retrieved from http://www.youcubed.org/category/making-group-work-equal/

Youcubed at Stanford University. (2015b). Moving colors. Stanford Graduate School of Education. Retrieved from http://www.youcubed.org/task/moving-colors/

Youcubed at Stanford University. (2015c). Tour of mathematical connections. Stanford Graduate School of Education. Retrieved from http://www.youcubed.org/tour-of-mathematical-connections/

Youcubed at Stanford University. (2015d). Video: High-quality teaching examples. Stanford Graduate School of Education. Retrieved from www.youcubed.org/high-quality-teaching-examples/

Zaleski, A. (2014, November 12). Western High School's RoboDoves crushes the competition, stereotypes. *Baltimore Sun*. Retrieved from http://www.baltimoresun.com/entertainment/bthesite/bs-b-1113-cover-robodoves-20141111-story.html

Zohar, A., & Sela, D. (2003). Her physics, his physics: Gender issues in Israeli advanced placement physics classes. *International Journal of Science Education, 25*(2), 261.

Appendix A

Math Homework Reflection Questions
Part 1: Written Response Questions

*Your response to the question(s) chosen should be very detailed! Please write in complete sentences and be ready to share your response in class the next day.

1. What were the main mathematical concepts or ideas that you learned today or that we discussed in class today?

2. What questions do you still have about _____? If you don't have a question, write a similar problem and solve it instead.

3. Describe a mistake or misconception that you or a classmate had in class today. What did you learn from this mistake or misconception?

4. How did you or your group approach today's problem or problem set? Was your approach successful? What did you learn from your approach?

5. Describe in detail how someone else in class approached a problem. How is their approach similar or different to the way you approached the problem?

6. What new vocabulary words or terms were introduced today? What do you believe each new word means? Give an example/picture of each word.

7. What was the big mathematical debate about in class today? What did you learn from the debate?

8. How is _____ similar or different to _____?

9. What would happen if you changed _____?

10. What were some of your strengths and weaknesses in this unit? What is your plan to improve in your areas of weakness?

Shapes Task

How do you see the shapes growing?

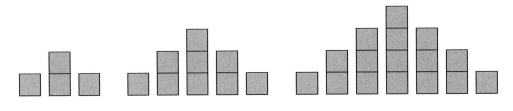

Source: From Ruth Parker; a task used in MEC courses.

Cuisenaire Rod Trains

Find out how many different trains can be made for any length of rod. For example, with the light green rod you can make these four trains:

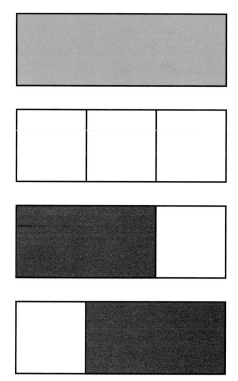

Source: From Ruth Parker; a task used in MEC courses.

Pascal's Triangle

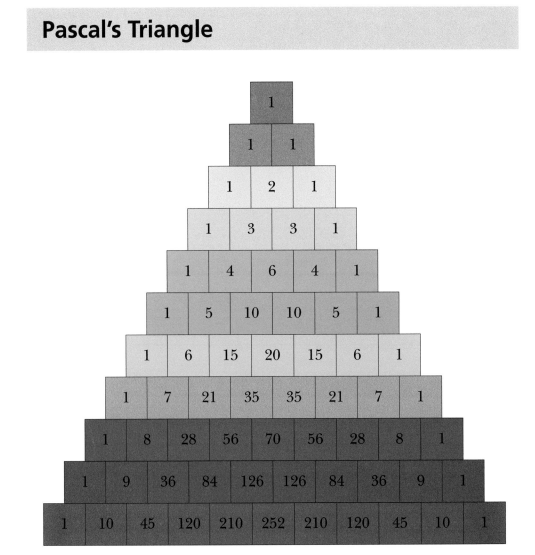

Negative Space Task

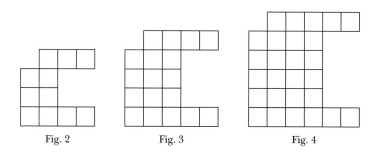

Fig. 2   Fig. 3   Fig. 4

1. What would figure 100 look like?

2. Imagine you could continue your pattern backward. How many tiles would there be in Figure –1? (That's figure negative one, whatever that means!)

3. What would Figure –1 look like?

Source: Adapted from Carlos Cabana.

Find Quadrilaterals!

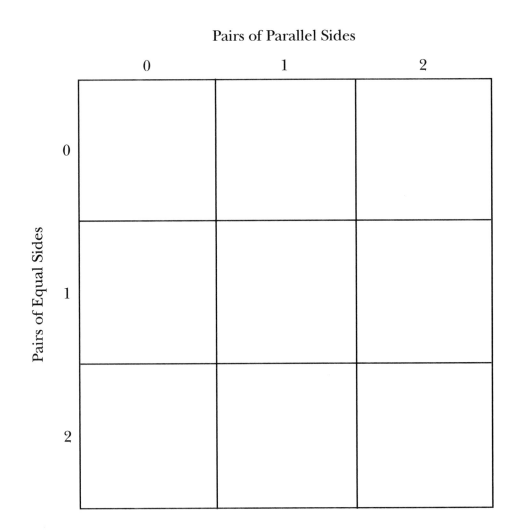

Pairs of Parallel Sides

	0	1	2
0			
1			
2			

Pairs of Equal Sides

Four 4's

Can you find every number between 1 and 20 using only four 4's and any operation?

Going beyond …

Can you find more than one way to make each number with four 4's?

Can you go beyond 20?

Can you use four 4's to find negative integers?

Newsletter

You are writing a newsletter to share your learning on this mathematics topic with your family and friends. You'll have the chance to describe your understanding of the ideas and write about why the mathematical ideas you have learned are important. You'll also describe a couple of activities that you worked on that were interesting to you.

In creating your newsletter, you can draw on the following resources:

- Photos of different activities
- Sketches
- Cartoons
- Interviews/surveys

To refresh your memory, here are some of the activities we've worked on:

Please prepare the following four sections. You can change the titles of the sections to fit your work.

Headline News	New Discoveries
Explain the big idea of the mathematics and what it means in at least two different ways. Use words, diagrams, pictures, numbers, and equations.	Choose at least two different activities from the work we have done that helped you understand the concepts. For each activity: • Explain why you chose the activity. • Explain what you learned about through the activity. • Explain what was challenging about the activity. • Explain the strategies you used to address your challenge.
Connections	**The Future**
Choose one additional activity that helped you learn a mathematical idea or process that you can connect to some other learning. • Explain why you chose the activity. • Explain the big mathematical idea you learned from the activity. • Explain what you connected this idea to and how you see the connection. • Explain the importance of the connection and how you might use this in the future.	Write a summary for the newsletter that addresses the following: • What is the big mathematical idea useful for? • What questions do you still have about the big idea?

The Long Jump

You are going to try out for the long jump team, for which you need an average jump of 5.2 meters. The coach says she will look at your best jump each day of the week and average them out. These are the five jumps you recorded that week:

	Meters
Monday	5.2
Tuesday	5.2
Wednesday	5.3
Thursday	5.4
Friday	4.4

Unfortunately, Friday's was a low score because you weren't feeling that well!

How could you work out an average that you think would fairly represent your jumping? Work out some averages in different ways and see which you think is most fair, then give an argument for why you think it is fairest. Explain your method and try and convince someone that your approach is best.

Parallel Lines and a Transversal

1. Use color coding to identify congruent angles.

2. Identify vertical and supplementary angles.

3. Write about the relationships you see. Use the color from your diagram in your writing.

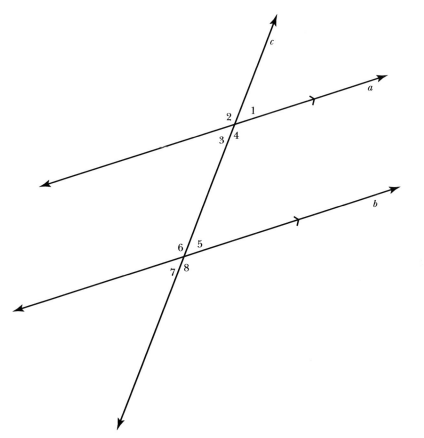

Vertical Angles:

Supplementary Angles:

Relationships:

Staircase

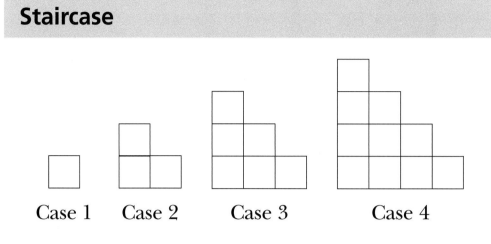

Case 1   Case 2   Case 3   Case 4

How do you see the pattern growing?

How many would be in the 100th case?

What about the nth case?

Paper Folding

Work with a partner. Take turns being the skeptic or the convincer. When you are the convincer, your job is to be convincing! Give reasons for all of your statements. Skeptics must be skeptical! Don't be easily convinced. Require reasons and justifications that make sense to you.

For each of the following problems, one person should make the shape and then be convincing. Your partner is the skeptic. When you move to the next question, switch roles.

Start with a square sheet of paper and make folds to construct a new shape. Then, explain how you know the shape you constructed has the specified area.

1. Construct a square with exactly 1/4 the area of the original square. Convince your partner that it is a square and has 1/4 of the area.

2. Construct a triangle with exactly 1/4 the area of the original square. Convince your partner that it has 1/4 of the area.

3. Construct another triangle, also with 1/4 the area, that is not congruent to the first one you constructed. Convince your partner that it has 1/4 of the area.

4. Construct a square with exactly 1/2 the area of the original square. Convince your partner that it is a square and has 1/2 of the area.

5. Construct another square, also with 1/2 the area, that is oriented differently from the one you constructed in 4. Convince your partner that it has 1/2 of the area.

Source: Adapted from Driscoll, 2007, p. 90, http://heinemann.com/products/E01148.aspx

Cone and Cylinder

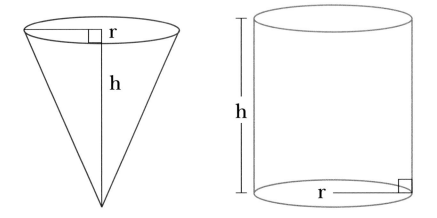

The height and radius of the cone and cylinder are the same. What is the relationship between the volume of the cone and the volume of the cylinder? Make a conjecture and try to convince other students. Use drawings, models, and color coding to be convincing.

My Homework
My Reflections

What was the main idea you learned
today?

I ♡ LOVE MATH

What is something you are struggling with or have questions
about?

How could the ideas from today's lesson be used in life?

Sorting the Numbers

Well, how about doing a simple jigsaw?

This problem has been designed to be worked on in a group of about four. (Some teacher notes and ideas for an extension are at http://nrich.maths.org/6947&part=note.)

1. There are two jigsaw puzzles that your teacher can print out for you (see below).

Complete each jigsaw and then put the pieces into the outline squares, which can be printed:

2. Place the smaller square of numbers on top of the other larger square in any way you like so that the small centimeter squares match up. (You may find it easier to copy the numbers on the smaller square onto a transparent sheet.)

3. Explore what happens when you add together the numbers that appear one on top of the other.

4. In your group, explore any other ideas that you come up with.

When you've looked at the 36 combinations, you probably need to ask, "I wonder what would happen if we …?" Change one small thing, explore that, and then compare your two sets of results.

You might like to ask, "Why …?"

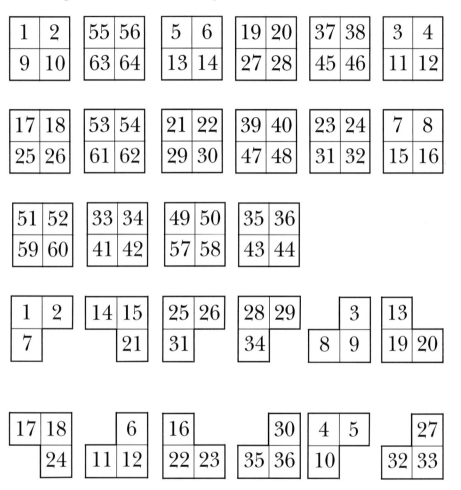

Source: From NRICH (http://nrich.maths.org/6947)

Growing Rectangles

Imagine a rectangle with an area of 20cm^2.

What could its length and width be? List at least five *different* combinations.

Imagine enlarging each of your rectangles by a scale factor of 2:

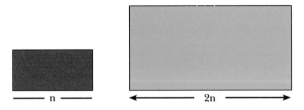

List the dimensions of your enlarged rectangles and work out their areas. What do you notice?

Try starting with rectangles with a different area and enlarge them by a scale factor of 2. What happens now?

Can you explain what's going on?

What happens to the area of a rectangle if you enlarge it by a scale factor of 3? Or 4? Or 5? What happens to the area of a rectangle if you enlarge it by a fractional scale factor?

What happens to the area of a rectangle if you enlarge it by a scale factor of k?

Explain and justify any conclusions you come to.

Do they apply to plane shapes other than rectangles?

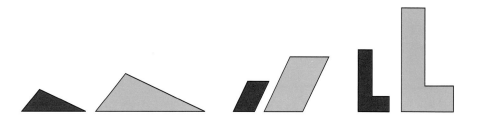

Now explore what happens to the surface area and volume of different cuboids when they are enlarged by different scale factors.

Explain and justify any conclusions you come to.

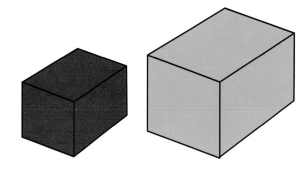

Do your conclusions apply to solids other than cuboids?

Source: From NRICH (http://nrich.maths.org/6923)

Linear Function Task

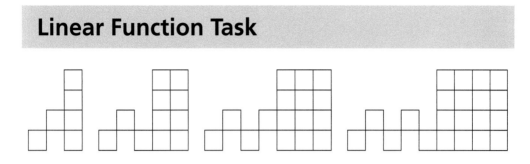

How do the shapes grow?

Can you predict what the 100th case would be?

What about the nth case?

Math Function Task

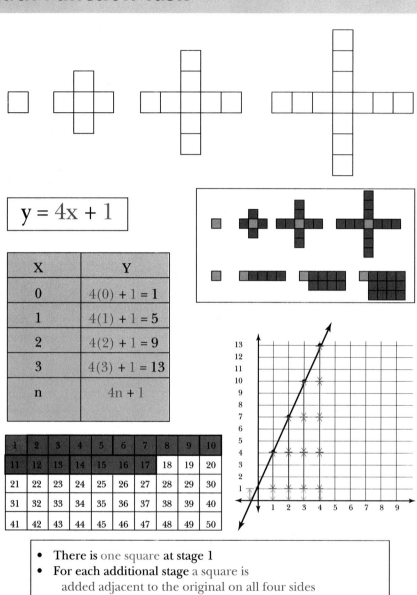

$$y = 4x + 1$$

X	Y
0	4(0) + 1 = 1
1	4(1) + 1 = 5
2	4(2) + 1 = 9
3	4(3) + 1 = 13
n	4n + 1

- There is one square at stage 1
- For each additional stage a square is added adjacent to the original on all four sides
- The figure continues to grow to the left, right, up and down, adding four squares for each new stage

Shoelaces

What length shoelaces are needed for different size shoes?

Investigate the relationship between shoelace length and shoe size.

Produce an equation in the form $y = mx + b$ that would help a shoemaker know the length of shoelaces they need to buy for different shoes.

Group Roles USA
Facilitator:

Make sure your group reads all the way through this card together before you begin. "Who wants to read? Does every one get what to do?"

Keep your group together. Make sure everyone's ideas are heard. "Did anyone see it a different way? Are we ready to move on?" Be sure everyone can explain.

Recorder/Reporter:

Your group needs to organize all your results. Your results need to show everyone's ideas, be well organized, and use color, arrows, and other math tools to communicate your mathematics, reasons, and connections. "How do we want to show that idea?" Be ready to join the teacher for a huddle.

Resource Manager:

- Get materials for your team.
- Make sure all questions are team questions.
- When your team is done, call the teacher over to debrief the mathematics.

Team Captain:

- Remind your team to find reasons for each mathematical statement and search for connections among the different statements. "How do you know that for sure? How does that relate to … ?"
- No talking outside your group!

Group Roles, British
Organiser:

- Keep the group together and focused on the problem; make sure no one is talking to people outside the group.

Resourcer:

- You are the only person that can leave their seat to collect rulers, calculators, pencils, etc., for the group.
- Make sure everyone is ready before you call the teacher.

Understander:

- Make sure all ideas are explained so everyone is happy with them.
- If you don't understand, ask whoever had the idea … if you do, make sure that everyone else does too.
- Make sure that all the important parts of your explanation get written down.

Includer:

- Make sure everyone's ideas are listened to; invite other people to make suggestions.

Self-Assessment: Polygons

	I can do this independently and explain my solution path(s) to my classmate or teacher.	I can do this independently.	I need more time. I need to see an example to help me.
Draw lines and line segments with given measurements.			
Draw parallel lines and line segments.			
Draw intersecting lines and line segments.			
Create a polygon with a given perimeter.			
Create a square or rectangle with a given area.			
Create an irregular shape whose area can be solved by cutting it into rectangles or squares.			

Source: From Lori Mallett

Algebra 1 Self-Assessment

Unit 1 – Linear Equations and Inequalities

☐ I can solve a linear equation in one variable.

☐ I can solve a linear inequality in one variable.

☐ I can solve formulas for a specified variable.

☐ I can solve an absolute value equation in one variable.

☐ I can solve and graph a compound inequality in one variable.

☐ I can solve an absolute value inequality in one variable.

Unit 2 – Representing Relationships Mathematically

☐ I can use and interpret units when solving formulas.

☐ I can perform unit conversions.

☐ I can identify parts of an expression.

☐ I can write the equation or inequality in one variable that best models the problem.

☐ I can write the equation in two variables that best model the problem.

☐ I can state the appropriate values that could be substituted into an equation and defend my choice.

☐ I can interpret solutions in the context of the situation modeled and decide if they are reasonable.

☐ I can graph equations on coordinate axes with appropriate labels and scales.

☐ I can verify that any point on a graph will result in a true equation when their coordinates are substituted into the equation.

☐ I can compare properties of two functions graphically, in table form, and algebraically.

Unit 3 – Understanding Functions

☐ I can determine if a graph, table, or set of ordered pairs represents a function.

☐ I can decode function notation and explain how the output of a function is matched to its input.

☐ I can convert a list of numbers (a sequence) into a function by making the whole numbers the inputs and the elements of the sequence the outputs.

☐ I can identify key features of a graph, such as the intercepts, whether the function is increasing or decreasing, maximum and minimum values, and end behavior, using a graph, a table, or an equation.

☐ I can explain how the domain and range of a function is represented in its graph.

Unit 4 – Linear Functions

☐ I can calculate and interpret the average rate of change of a function.

☐ I can graph a linear function and identify its intercepts.

☐ I can graph a linear inequality on a coordinate plane.

☐ I can demonstrate that a linear function has a constant rate of change.

☐ I can identify situations that display equal rates of change over equal intervals and can be modeled with linear functions.

☐ I can construct linear functions from an arithmetic sequence, graph, table of values, or description of the relationship.

☐ I can explain the meaning (using appropriate units) of the slope of a line, the y-intercept, and other points on the line when the line models a real-world relationship.

Unit 5 – Systems of Linear Equations and Inequalities

☐ I can solve a system of linear equations by graphing.

☐ I can solve a system of linear equations by substitution.

☐ I can solve a system of linear equations by the elimination method.

☐ I can solve a system of linear inequalities by graphing.

☐ I can write and graph a set of constraints for a linear-programming problem and find the maximum and/or minimum values.

Unit 6 – Statistical Models

☐ I can describe the center of the data distribution (mean or median).

☐ I can describe the spread of the data distribution (interquartile range or standard deviation).

☐ I can represent data with plots on the real number line (dot plots, histograms, and box plots).

☐ I can compare the distribution of two or more data sets by examining their shapes, centers, and spreads when drawn on the same scale.

☐ I can interpret the differences in the shape, center, and spread of a data set in the context of a problem, and can account for effects of extreme data points.

☐ I can read and interpret the data displayed in a two-way frequency table.

☐ I can interpret and explain the meaning of relative frequencies in the context of a problem.

☐ I can construct a scatter plot, sketch a line of best fit, and write the equation of that line.

☐ I can use the function of best fit to make predictions.

☐ I can analyze the residual plot to determine whether the function is an appropriate fit.

☐ I can calculate, using technology, and interpret a correlation coefficient.

☐ I can recognize that correlation does not imply causation and that causation is not illustrated on a scatter plot.

Unit 7 – Polynomial Expressions and Functions

☐ I can add and subtract polynomials.

☐ I can multiply polynomials.

☐ I can rewrite an expression using factoring.

☐ I can solve quadratic equations by factoring.

☐ I can sketch a rough graph using the zeroes of a quadratic function and other easily identifiable points.

Unit 8 – Quadratic Functions

☐ I can use completing the square to rewrite a quadratic expression into vertex form.

☐ I can graph a quadratic function, identifying key features such as the intercepts, maximum and/or minimum values, symmetry, and end behavior of the graph.

☐ I can identify the effect of transformations on the graph of a function with and without technology.

☐ I can construct a scatter plot, use technology to find a quadratic function of best fit, and use that function to make predictions.

Unit 9 – Quadratic Equations

☐ I can explain why sums and products are either rational or irrational.

☐ I can solve quadratic equations by completing the square.

☐ I can solve quadratic equations by finding square roots.

☐ I can solve quadratic equations by using the quadratic formula.

Unit 10 – Relationships That Are Not Linear

☐ I can apply the properties of exponents to simplify algebraic expressions with rational exponents.

☐ I can graph a square root or cube root function, identifying key features such as the intercepts, maximum and/or minimum values, and end behavior of the graph.

☐ I can graph a piecewise function, including step and absolute value functions, identifying key features such as the intercepts, maximum and/or minimum values, and end behavior of the graph.

Unit 11 – Exponential Functions and Equations

☐ I can demonstrate that an exponential function has a constant multiplier over equal intervals.

☐ I can identify situations that display equal ratios of change over equal intervals and can be modeled with exponential functions.

☐ I can use graphs or tables to compare the rates of change of linear, quadratic, and exponential functions.

☐ I can rewrite exponential functions using the properties of exponents.

☐ I can interpret the parameters of an exponential function in real-life problems.

☐ I can graph exponential functions, identifying key features such as the intercepts, maximum and/or minimum values, asymptotes, and end behavior of the graph.

☐ I can construct a scatter plot, use technology to find an exponential function of best fit, and use that function to make predictions.

Source: By Lisa Henry

Two Stars and a Wish

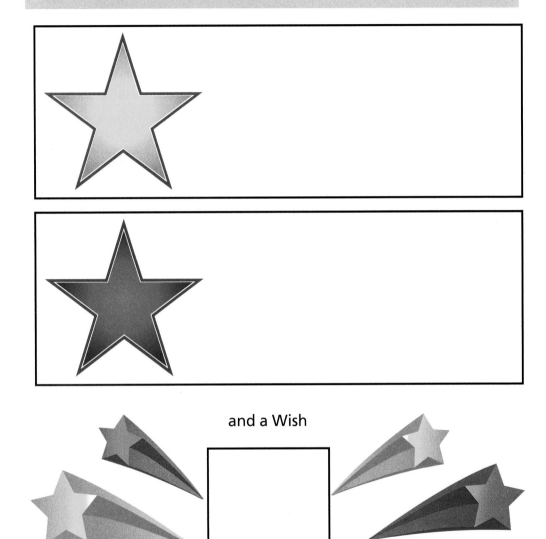

and a Wish

Reflection

What was the big idea we worked on today?

What did I learn today?

What good ideas did I have today?

In what situations could I use the knowledge I learned today?

What questions do I have about today's work?

What new ideas do I have that this lesson made me think about?

Algebra Jigsaw Task A

A Task

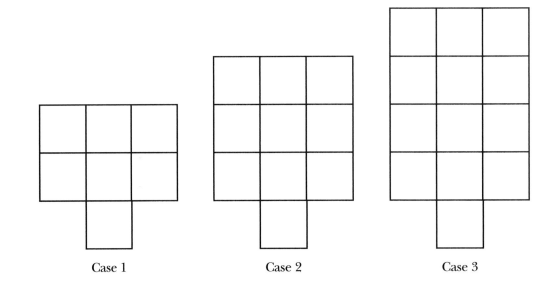

Case 1   Case 2   Case 3

Algebra Jigsaw Task B

B Task

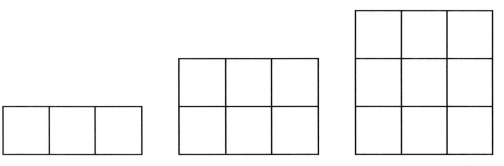

Case 1   Case 2   Case 3

Algebra Jigsaw Task C

C Task

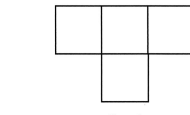
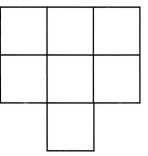

Case 1   Case 2   Case 3

Jigsaw Algebra Task D

D Task

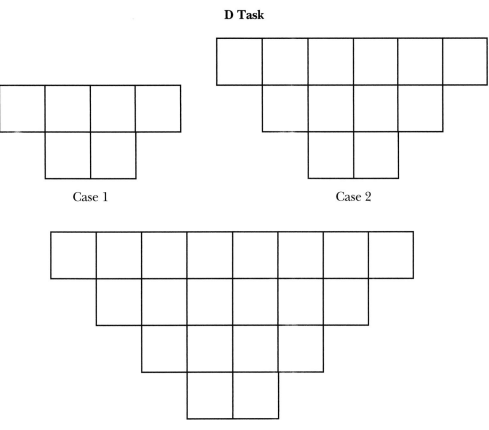

Case 1   Case 2

Case 3

Exit Ticket

Exit Ticket		Name Date
Three things I learned today …	Two things I found interesting …	One question I have …

Show What You Can Do Self-Assessment

What we value from an individual	Justify (if necessary)	
Perseverance • Did you stick with it? • Did you try something else? • Did you ask a question? • Did you describe where you're stuck?		Did it! Approved
Multiple Representations Words Pictures Charts Diagrams Graphs More than one solution process Data Table		Did it! Approved
Clear Expectations • Did you describe your thinking process? • How did you get your answer? *or* Where did you get stuck? • Ideas: arrows, color, words, numbers		Did it! Approved
Product • Did you complete the task, or where did you get stuck? • Did you give the task your best effort?		Did it! Approved

Source: From Ellen Crews.

Participation Quiz Mathematical Goals

Your group will be successful today if you are ...

- Recognizing and describing patterns
- Justifying thinking and using multiple representations
- Making connections between different approaches and representations
- Using words, arrows, numbers, and color coding to communicate ideas clearly
- Explaining ideas clearly to team members and the teacher
- Asking questions to understand the thinking of other team members
- Asking questions that push the group to go deeper
- Organizing a presentation so that people outside the group can understand your group's thinking

No one is good at all of these things, but everyone is good at something. You will need all of your group members to be successful at today's task.

Source: From Carlos Cabana.

Participation Quiz Group Goals

During the participation quiz, I will be looking for ...

- Leaning in and working in the middle of the table
- Equal air time
- Sticking together
- Listening to each other
- Asking each other lots of questions
- Following your team roles

Source: From Carlos Cabana.

Dog Biscuits

How many ways can you make two groups of 24 dog biscuits?

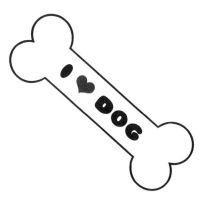

How many ways can you equally group 24 dog biscuits?

Show your results in a visual representation that shows all of the combinations.

Highlighting Mathematical Connections

Show the fractions 3/4, 6/8, and 12/16 on a graph.

Show the fractions as similar triangles.

What is similar and different about the different fraction representations—as numbers, a graph, and triangles? Can you color code features of each representation so that they show up in the same color on the different representations?

Color Coding Brownies

Sam has made a pan of brownies that he wants to cut into 24 equal pieces. He wants to share them equally with 5 of his friends. Partition the pan of brownies and use color coding to show how many Sam and his friends will get.

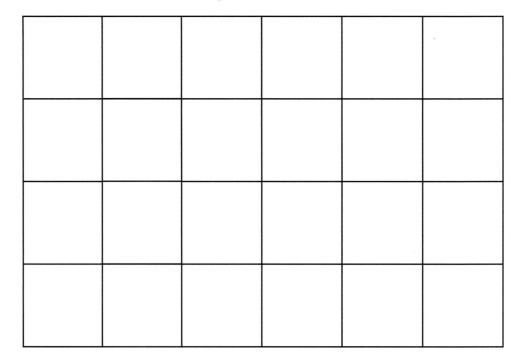

Painted Cube

Imagine a 5 × 5 × 5 cube that had been painted blue on the outside, with cubes made up of smaller 1 × 1 × 1 cubes. Consider the questions:

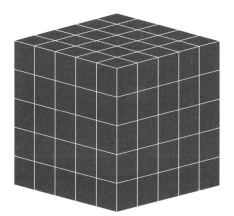

How many small cubes will have 3 blue faces?

How many small cubes will have 2 blue faces?

How many small cubes will have 1 blue face?

How many small cubes will have no paint on them?

The Tethered Goat

Imagine a goat tied to the corner of a shed by a rope. The shed is 4 feet by 6 feet. The rope is 6 feet long.

What do you wonder about this situation?

Draw a picture of the situation.

What questions do you have?

The sun rises to the east of the shed and sets to the west. The goat would appreciate some shade. Where should you plant a tree? What tree would you plant?

World Wealth Simulation

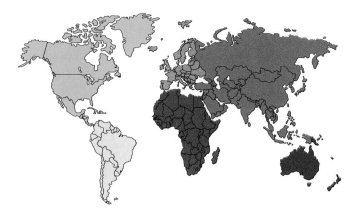

1. Find the percentage of the world's population living on each continent.

2. Calculate the number of people in our class who would correspond to the percentages found.

3. Calculate the percentage of the world's wealth for each continent.

4. Calculate the wealth of each continent in cookies.

TABLE 1 World Wealth Data

Continent	Population (in millions) 2000	Percent of Population	Wealth (GDP in trillions of dollars)	Percentage of Wealth
Africa	1,136		2.6	
Asia	4,351		18.5	
N. America	353		20.3	
S. America	410		4.2	
Europe	741		24.4	
Oceania/Aust.	39		1.8	
Total	7,030	100%	71.8	100%

Sources: Population data according to Population Reference Bureau (prb.org). Wealth data according to International Monetary Fund.

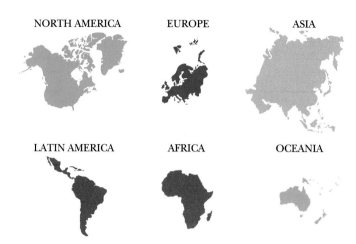

NORTH AMERICA   EUROPE   ASIA

LATIN AMERICA   AFRICA   OCEANIA

TABLE 2   Classroom Data

Number of people in classroom _____

Total number of cookies _____

Continent	Pop %	# of People in Class	Wealth %	# Cookies
Africa				
Asia				
N. America				
S. America				
Europe				
Oceania/Aust.				
Total	~100%		~100%	

Source: From Charmaine Mangram.

Soccer Goalie

If you are a soccer goalie, and an attacking player from the other team has broken away from the others and is running toward you, where is the best place for you to stand? Try mapping out different positions depending on the location of the attacking player when she shoots.

We Wonder

Team Members:

Date:

We wonder

_____

_____

Use pictures, numbers, and words to show how you answered your question.

We want to investigate

_____

_____

Use pictures, numbers, and words to show how you answered your question.

Source: From Nick Foote.

Appendix B: Setting up Positive Norms in Math Class

By Jo Boaler

Here are 7 of my favorite messages to give to students in math class, and some suggestions from *you*cubed as to how to encourage them:

Everyone can learn math to the highest levels

Mistakes are valuable

Questions are really important

Math is about creativity and making sense

Math is about connections and communicating

Math class is about learning not performing

Depth is more important than speed

1. Everyone can learn math to the highest levels.

Encourage students to believe in themselves. There are different parts to this – first we need students to know that they can achieve at any math level, and there is no such thing as a math person. Brain information is really good for this.

Second we need them to have a "growth mindset" – believing that they can learn anything, and the more work they do the smarter they will get.

An important way to encourage a growth mindset is by praising what students have done and learned, not them as a person. So instead of saying "you are so smart", say "it is great that you have learned that."

Some videos you might want to share with students to encourage positive brain messages and a growth mindset:

youcubed.org/teachers/from-stanford-onlines-how-to-learn-math-for-teachers-and-parents-brain-science

youcubed.org/students/boosting-messages

What is a growth mindset?

There is a really damaging myth that pervades the US/UK and other countries – the idea that some people are born with a "math brain" and some are not. This has been resoundingly disproved by research but many students and parents believe this. It is really important to communicate "growth mindset" messages to students. Help them know that everyone is a math person and that the latest research is telling us that students can reach any levels in math because of the incredible plasticity of the brain.

2. Mistakes are valuable

Tell students that you love mistakes and that they will be valued at all times, tell them that it is good to make mistakes as we know that when people make mistakes, their brains are growing. This single message can be incredibly liberating for students. Here are some suggestions for encouraging positive thinking about mistakes:

1. Ask students with mistakes to present mistakes (especially deep, conceptual ones) on the board so that everyone can learn from them. If one student makes a conceptual mistake, there are probably many others making the same one.

2. When students get something wrong – instead of being discouraging or sympathetic, say "your brain just grew! Synapses are firing, that's really good"

3. Ask students to read positive brain/mistake messages and choose their favorites that they will take on for the year. Eg "easy is a waste of time" "working hard grows your brain" "it is really important to make mistakes". Ask them to draw brains with the messages on them that you can display on your walls, see right.

4. Crumpled Paper: Ask students to crumple a piece of paper and throw it at the board with the feeling they have when making a mistake. Then get them to retrieve the paper and color in all the lines, these represent synapses firing and brain growth from making a mistake. Ask them to keep the piece of paper in their math folders/notebooks to remind them of this.

Research shows that when students make mistakes, synapses fire and brains grow. Brain activity is particularly strong in individuals with a growth mindset. It is good to make mistakes.

Activity 3 from Kim Hollowell at Vista Unified.

3. Questions are really important.

Tell your students that you love questions about math and that they are really important. Research shows us that question asking is linked to high achievement – yet as students move through school they ask fewer and fewer questions, for fear of being thought clueless. You don't need to be able to answer every question that students may come up with, sometimes it is good to say that you don't know but you will find out, or ask other students if someone would like to answer the question.

Some suggestions for encouraging questions:

1. When good questions are asked, write them in large colored letters onto posters that you post around the room, to celebrate them. Show questions from a range of students.

2. Tell students they have 2 responsibilities in your classroom. One is to always ask a question if they have one, and the other it to always answer a question from classmates if asked.

3. Encourage students to ask questions – from you, other students and themselves, such as: why does that work? why does that make sense? Can I draw that? How does that method connect to another?

4. Encourage students to ask their own math questions. Instead of asking questions for them, give them interesting mathematical situations and see what questions arise for them.

In studies, student question asking has been shown to steadily decline as students go through the grades in the US, showing this relationship:

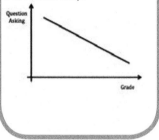

4. Math is about creativity and making sense.

The key to understanding math is making sense of it. Many students believe that math is a set of formulas that have to be remembered - this belief is associated with low achievement. Math is a very creative subject that is, at its core, about visualizing patterns and creating solution paths that others can see, discuss and critique.

Some methods for encouraging sense making and creative math:

PISA data from 15 million 15-year olds worldwide shows that the lowest achieving students in the world are those who believe that mathematical success comes from memorization. The USA and UK are countries where the highest numbers of students believe this.

1. Always ask students – why does that make sense? Ask this whether their answers are correct or incorrect

2. Encourage visual mathematics. Ask students to draw their solutions. Ask them to think about how they see math. In this video (http://youtu.be/1EqrX-gsSQg) Cathy Humphreys asks students to make sense of 1 divided by 2/3 by drawing their solutions.

3. Show mathematical ideas through visual representations. All mathematics can be represented visually, and visual representations give many more students access to understanding. We have many examples of visual mathematics on youcubed and in the classroom video above.

4. Use number talks that value students' different ways of seeing math and solving problems. This video teaching number talks also shows visual solutions.
http://youcubed.org/teachers/2014/from-stanford-onlines-how-to-learn-math-for-teachers-and-parents-number-talks/
5. When students finish questions, ask them to think of new, harder questions. These could be questions to give to other students. This is a really good strategy for differentiation.

5. Math is about connections and communicating.

Math is a connected subject, but students often think it is a set of disconnected methods. We made a video to show some connections and students loved it.
youcubed.org/students/a-tour-of-mathematical-connections

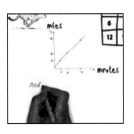

Mathematics is a form of communication, some people think of it as a language. Some strategies for encouraging connecting and communicating are:

1. Show the connections video.
2. Encourage students to represent their math results in different forms eg words, a picture, a graph, an equation, and to link between them, see below.
3. Encourage color coding, ask students to show with color where a mathematical idea is, see below.

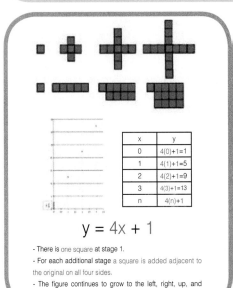

x	y
0	4(0)+1=1
1	4(1)+1=5
2	4(2)+1=9
3	4(3)+1=13
n	4(n)+1

$$y = 4x + 1$$

- There is one square at stage 1.
- For each additional stage a square is added adjacent to the original on all four sides.
- The figure continues to grow to the left, right, up, and down, adding four squares for each new stage.

1	2	3	4	5	6	7	8	9	10
11	12	13	14	15	16	17	18	19	20
21	22	23	24	25	26	27	28	29	30
31	32	33	34	35	36	37	38	39	40
41	42	43	44	45	46	47	48	49	50
51	52	53	54	55	56	57	58	59	60
61	62	63	64	65	66	67	68	69	70
71	72	73	74	75	76	77	78	79	80
81	82	83	84	85	86	87	88	89	90
91	92	93	94	95	96	97	98	99	100

6. Value depth over speed.

Many people incorrectly believe that being good at math means being fast at math. It doesn't and we need to dissociate math from speed. When we value fast computation (as many classrooms do) we encourage a subset of learners who compute quickly and discourage many others, including deep slow thinkers who are very important to math (see sidebar).

We no longer need students to compute fast (we have computers for this) we need them to think deeply, connect methods, reason, and justify.

1. Tell students you don't value fast work. Mathematical thinking is about depth not speed.
2. Don't let mathematical discussions be driven by the fastest students.
3. When asking for hands up, don't always take answers from the fastest students.
4. Don't use flash cards, speed competitions, timed tests, instead value depth, creativity, different ways of thinking about math, and different explanations. A paper showing the research suggesting timed tests cause math anxiety is here: http://youcubed.org/pdfs/nctm-timed-tests.pdf

Research suggests that timed tests cause math anxiety *

"I was always deeply uncertain about my own intellectual capacity; I thought I was unintelligent And it is true that I was, and still am, rather slow. I need time to seize things because I always need to understand them fully. Towards the end of the eleventh grade, I secretly thought of myself as stupid. I worried about this for a long time.

I'm still just as slow. (...)At the end of the eleventh grade, I took the measure of the situation, and came to the conclusion that rapidity doesn't have a precise relation to intelligence. What is important is to deeply understand things and their relations to each other. This is where intelligence lies. The fact of being quick or slow isn't really relevant."

- Laurent Schwartz,
Winner of the Fields Medal
 (A Mathematician Grappling with His Century, 2001)

7. Math class is about learning, not performing.

Many students think that their role in math class is not to learn but to get questions right – to perform. It is important for them to know that math is about learning, and to know that math is a growth subject, it takes time to learn and it is all about effort. Some strategies for making math a learning, not a performing subject:

1. Grade and test less. Math is the most over-graded, over-tested subject in the curriculum. Neither grades nor tests have been shown to increase learning, from research, and both make students feel they are performing and not learning. Grades often make students think they are a reflection not of what they have learned but who they are. There is a video reflecting this at http://youtu.be/eoVLBExuqB0

2. Instead, give diagnostic comments. These take longer but are extremely valuable and can be done less often.

3. Use "assessment for learning" strategies (see sidebar).

4. If you have to grade, then give grades for learning, not for performing eg for asking questions, representing ideas in different ways, explaining work to others, making connections. Assess the breadth of math, not just a small part of math – procedure execution.

5. You may have to give grades to your administration but that doesn't mean you have to give them to the students. Grades communicate fixed messages about learning and are often counter-productive for students.

Assessment for learning (A4L) teaching strategies have been shown to drastically increase student achievement, if they are used instead of summative tests and grades. It has been estimated that if teachers in England used A4L strategies the achievement of their students would increase so much the country would move, in international comparisons, from the middle of the pack to the top 5 (Black and Wiliam, 1998). At https://www.youcubed.org/category/assessment-and-grading/ we are sharing our favorite A4L strategies.

Positive Norms to Encourage in
Math Class   By Jo Boaler

1. Everyone Can Learn Math to the Highest Levels.
Encourage students to believe in themselves. There is no such thing as a "math" person. Everyone can reach the highest levels they want to, with hard work.

2. Mistakes are valuable
Mistakes grow your brain! It is good to struggle and make mistakes.

3. Questions are Really Important
Always ask questions, always answer questions. Ask yourself: why does that make sense?

4. Math is about Creativity and Making Sense.
Math is a very creative subject that is, at its core, about visualizing patterns and creating solution paths that others can see, discuss and critique.

5. Math is about Connections and Communicating
Math is a connected subject, and a form of communication. Represent math in different forms eg words, a picture, a graph, an equation, and link them. Color code!

6. Depth is much more important than speed.
Top mathematicians, such as Laurent Schwartz, think slowly and deeply.

7. Math Class is about Learning not Performing
Math is a growth subject, it takes time to learn and it is all about effort.

ABOUT THE AUTHOR

Dr. Jo Boaler is a Professor of Mathematics Education at Stanford University and the co-founder of Youcubed. She is also an analyst for PISA testing in the OECD, and author of the first MOOC on mathematics teaching and learning. Formerly she was the Marie Curie Professor of Mathematics Education in England. She won the award for the best PhD in England, from the British Educational Research Association. She is an elected fellow of the Royal Society of Arts (Great Britain), and a former president of the International Organization for Women and Mathematics Education (IOWME). She is the recipient of a National Science Foundation Presidential Award and the NCSM Kay Gilliland Equity Award. She is the author of nine books and numerous research articles. She serves as an advisor to several Silicon Valley companies and as a White House presenter on girls and STEM. She recently formed www.youcubed.org to give teachers and parents the resources and ideas they need to inspire and excite students about mathematics.

ACKNOWLEDGMENTS

This book could only have been written with the support, collaboration, and ideas of the teachers that I work with, connect with through social media, and learn from in my role as a professor of mathematics education. I am so fortunate to know teachers who inspire students on a daily basis with their teaching and words and who invite me into their classrooms and lives. It is they who allow me to take research ideas and translate them into teaching ideas. This book shares some of the work of wonderful K–12 teachers, some of whom I introduce in the pages of this book. I am extremely grateful to all the teachers with whom I connect and work. We work together, as Youcubians, to share good practice and to learn from each other constantly. This is an enriching and fulfilling experience and this book is the outcome of our work together.

I also want to thank my two daughters, to whom I dedicated this book, for being so patient when I lock myself away to write and when I disappear for days on end to cross the country to work with other educators. At the time I am writing this, they are nine and twelve and they are my greatest inspiration, every day.

The fact that this book exists is due in no small part to the most patient book editor I have ever met or probably will ever meet! Kate Bradford has waited for years for me to finish this book. Whenever I wrote to Kate, guiltily asking for yet another extension, she shocked me with her kindness and patience. I hope, Kate, that this will be worth the wait!

I would also like to thank Jill Marsal, my agent, whose steadfast advice and encouragement has helped me from book inception to production.

And last, but not least, I want to thank my Youcubed partner and co-founder, Cathy Williams. Cathy produced nearly all the visuals for the book, discussed my half formed ideas with me, read drafts, and has been my biggest help and supporter all along the way. Everybody should have a Cathy in their life! Thank you Cathy for being a great advisor and friend.

Index